Girls and Computers

Bedford Way Papers

ISSN 0261—0078

1. 'Fifteen Thousand Hours': A Discussion
Barbara Tizard *et al.*
ISBN 0 85473 090 7

3. Issues in Music Education
Charles Plummeridge *et al.*
ISBN 0 85473 105 9

4. No Minister: A Critique of the D E S Paper 'The School Curriculum'
John White *et al.*
ISBN 0 85473 115 6

5. Publishing School Examination Results: A Discussion
Ian Plewis *et al.*
ISBN 0 85473 116 4

8. Girls and Mathematics: The Early Years
Rosie Walden and Valerie Walkerdine
ISBN 0 85473 124 5

9. Reorganisation of Secondary Education in Manchester
Dudley Fiske
ISBN 0 85473 125 3

11. The Language Monitors
Harold Rosen
ISBN 0 85473 134 2

12. Meeting Special Educational Needs: The 1981 Act and its Implications
John Welton *et al.*
ISBN 0 85473 136 9

13. Geography in Education Now
Norman Graves *et al.*
ISBN 0 85473 219 5

14. Art and Design Education: Heritage and Prospect
Anthony Dyson *et al.*
ISBN 0 85473 245 4

15. Is Teaching a Profession?
Peter Gordon (ed.)
ISBN 0 85473 220 9

16. Teaching Political Literacy
Alex Porter (ed.)
ISBN 0 85473 154 7

17. Opening Moves: Study of Children's Language Development
Margaret Meek (ed.)
ISBN 0 85473 161 X

18. Secondary School Examinations
Jo Mortimore, Peter Mortimore and Clyde Chitty
ISBN 0 85473 259 4

19. Lessons Before Midnight: Educating for Reason in Nuclear Matters
The Bishop of Salisbury *et al*
ISBN 0 85473 189 X

20. Education plc?: Headteachers and the New Training Initiative
Janet Maw *et al*
ISBN 0 85473 191 1

21. The Tightening Grip: Growth of Central Control of the School Curriculum
Denis Lawton
ISBN 0 85473 201 2

22. The Quality Controllers: A Critique of the White Paper 'Teaching Quality'
Frances Slater (ed.)
ISBN 0 85473 212 8

23. Education: Time for a New Act?
Richard Aldrich and Patricia Leighton
ISBN 0 85473 217 9

24. Girls and Mathematics: from Primary to Secondary Schooling
Rosie Walden and Valerie Walkerdine
ISBN 0 85473 222 5

25. Psychology and Schooling: What's the Matter?
Guy Claxton *et al.*
ISBN 0 85473 228 4

26. Sarah's Letters: A Case of Shyness
Bernard T. Harrison
ISBN 0 85473 241 1

27. The Infant School: past, present and future
Rosemary Davis (ed.)
ISBN 0 85473 250 0

28. The Politics of Health Information
Wendy Farrant and Jill Russell
ISBN 0 85473 260 8

29. The GCSE: an uncommon examination
Caroline Gipps (ed.)
ISBN 0 85473 262 4

30. Education for a Pluralist Society
Graham Haydon (ed.)
ISBN 0 85473 263 2

31. Lessons in Partnership: an INSET course as a case study
Elizabeth Cowne and Brahm Norwich
ISBN 0 85473 267 5

32. Redefining the Comprehensive Experience
Clyde Chitty (ed.)
ISBN 0 85473 280 2

33. The National Curriculum
Denis Lawton and Clyde Chitty (eds.)
ISBN 0 85473 294 2

34. Girls and Computers: general issues and case studies of Logo in the mathematics classroom
Celia Hoyles (ed.)
ISBN 0 85473 306 X

35. Training for School Management: Lessons from the American experience
Bruce S. Cooper and R. Wayne Shute
ISBN 0 85473 307 8

Girls and Computers

General Issues and Case Studies of Logo in the Mathematics Classroom

Edited by Celia Hoyles

Contributors:
Ann Brackenbridge, Alan Bibby, Jeremy Burke, Susan Burns, Julie-Ann Edwards, Pam Greenhaugh, Celia Hoyles, Martin Hughes, John Jeffries, Keith Jones, Judi Miln, Maggie Montgomery, Richard Noss, Bridget Perkins, Andrew Seager, Teresa Smart, Rosamund Sutherland, Helen Wright

Bedford Way Papers 34
INSTITUTE OF EDUCATION
University of London

First published in 1988 by the Institute of Education, University of London,
20 Bedford Way, London WC1H 0AL.

Distributed by Turnaround Distribution Ltd., 27 Horsell Road,
London N5 1XL (telephone: 01-609 7836).

British Library Cataloguing in Publication Data

Girls and Computers
1. Great Britain. Schools. Curriculum subjects : Mathematics. Teaching.
Applications of microcomputer systems.
I. Hoyles, Celia, *1946*— II. Series.
510'.7'8

ISBN 0-85473-306-X

Typeset by Acies Print Ltd, Leicester
Printed in Great Britain by Billing & Sons Ltd, Worcester and London

Contents

Notes on Contributors

Celia Hoyles has been Professor of Mathematics Education at the Institute of Education, University of London, since 1984. Her main research interests concern the design of computer-based microworlds for mathematics, the analysis of peer interaction and group processes, and the development and evaluation of INSET programmes. Her book (with Rosamund Sutherland) *Logo Mathematics in the Classroom* is to be published early in 1989.

Alan Bibby is Head of Mathematics at Axminster Grammar School. He is currently studying towards an MEd in Educational Computing at the School of Education, University of Exeter.

Ann Brackenbridge is Research Assistant at the School of Education, University of Exeter. She has worked on research projects involving children's use of computers and their attitudes towards them.

Susan Burns is Research Fellow at King's College, University of London. She has recently completed an MSc in mathematics and computing education at the Institute of Education, University of London.

Pam Greenhaugh teaches at Foxhayes First School, Exeter, where she has responsibility for maths and computing. She was recently research assistant on the Nuffield-funded project 'Gender and Computing in Primary School Children', and is currently studying towards an MEd in Educational Computing at the School of Education, University of Exeter.

Martin Hughes is Lecturer in Education at the University of Exeter. He has carried out research into several aspects of the thinking and learning of young children. His books include *Young Children Learning* (with Barbara Tizard) and *Children and Number*.

Richard Noss is Senior Lecturer in Mathematics Education at the Institute of Education, University of London. His interests include the role of computers in learning and teaching mathematics, and the social, cultural and political context of mathematics education.

Teresa Smart at the time of writing, was Co-ordinator of Mathematics at the Islington Sixth Form Centre, Inner London Education Authority. She is now Senior Lecturer in Mathematics Education at the Polytechnic of North London. She is currently working towards an MPhil in mathematics education (using Logo) at the Institute of Education, University of London.

Rosamund Sutherland is currently joint director (with Celia Hoyles and Richard Noss) of research projects on a computer-based programme for secondary mathematics and (with Celia Hoyles) on the role of peer-group discussion in a computer environment, at the Institute of Education, University of London.

The SMILE Teachers are: Jeremy Burke (Langdon Park School, Inner London Education Authority), Julie-Ann Edwards (St Thomas More School, Chelsea, ILEA), John Jeffries (Quintin Kynaston School, ILEA), Keith Jones (The Archbishop Michael Ramsey School ILEA), Judi Miln (Brondesbury and Kilburn School, Brent), Maggie Montgomery (Langdon Park School, ILEA), Bridget Perkins (Quintin Kynaston School, ILEA), Andrew Seager (Northumberland Park School, Haringey), Helen Wright (SMILE Centre, ILEA).

Introduction

Celia Hoyles

It is a matter of grave concern that our culture is defining computers as pre-eminently male machines. Despite the fact that in everyday life computers are becoming ubiquitous, the use of the computer in education seems to be following the traditional lines of gender bias in society. The present situation raises distinctly familiar questions of equity in terms of access to and use of technology. While girls and boys might show a similar appreciation of the significance computers might have for their personal futures, boys tend to be more positively disposed than girls towards computers, are more likely than girls to take optional computer courses in school, to report more frequent home use of computers, and tend to dominate the limited computer resources that are available in school. It is also the case that even when girls are able to obtain access to the machines in school, only a restricted set of activities (which exclude programming for example) are often deemed to be appropriate for them. Finally, few girls take up any employment using computer skills (other than data processing or word processing).

It is, therefore, difficult to avoid the disturbing conclusion that girls are learning less than boys about computers and therefore acquiring less understanding as to how they might use computers for their own purposes. A basic premise underlying this Bedford Way Paper is that gender differences in attitude to and competence with computers are neither inevitable nor immutable. Differences which have emerged arise from a variety of sources. First they arise from the social image of computers. This is an example of how deeply held 'default assumptions of which we are hardly aware, permeate our mental representations. Such assumptions are illustrated by Hofstadter (1985), in the following story.

A father and his son were driving to a ball game when their car stalled on the railroad tracks. In the distance a train whistle blew a warning. Frantically, the father tried to start the engine, but in his panic he couldn't turn the key and the car was hit by the onrushing train. An ambulance sped to the scene and picked them up. On the way to the hospital, the father died. The son was still alive but his condition was very serious, and he needed immediate surgery. The moment they arrived at the hospital, he was wheeled into an emergency operating room, and the surgeon came in, expecting a routine case. However, on seeing the boy, the surgeon blanched and muttered, 'I can't operate on this boy — he's my son.'

Hofstadter points out that, to most of us, 'bizarre worlds with such things as reincarnation come more easily to mind that the idea that a surgeon could be a woman!' (Hofstadter, 1985 p.139). In a similar way, the control of computers tends to call up a default image of a man rather than a woman. Differences between males' and females' relationship to computers also relate to the particular uses to which computers have been put; for example, the treatment of a computer merely as a 'topic' or object of knowledge has serious educational consequences for girls (see the Review of the Literature chapter, p.8). This leads on to the general question of access to computers and the time made available for computer experience. When there is competition over scarce resources, girls tend to 'lose out' and this has certainly been observed in relation to computers in schools. However, analysis that only examines forces that push girls away from the computers oversimplifies gender inequality issues. We must look beyond an essentially transitional situation dominated by shortage of hardware and software. *If it is true that experience with computers in school is crucial, then we must consider what type of experiences should be made available and how they should be organized.* There is a need for attention to be paid to the cognitive style associated with computer use, the type of software used and the ways computers have been organized and supported in classrooms.

What we look towards is a situation when the computer is another resource to be used when appropriate and set aside when not. As in Papert's description, we should use a computer like a pencil: 'we don't say when asked what we are doing "we are working with a pencil", we describe what we are using the pencil for!' The analogy of a computer as a pencil is not, however, completely accurate: the computer is *more* than a pencil and is potentially a powerful resource which can provoke us to reconstruct and reformulate our ideas. Paradoxically, when using the computer in this way, we have to think of it, at least to a certain extent, as an object of knowledge, *but* being used for a purpose. There is thus a need for a more general conception of the computer as an aid in the acquisition of knowledge alongside an acknowledgement that it can provide a new way of thinking about knowledge. All these uses are touched upon in the chapters of this volume. Many authors point to the importance of incorporating computer use into classroom practice and describe how this has been successfully undertaken. Generally it is posited that we must move to a situation in which teachers plan and develop *their own* ideas as to how to use the computer. The teachers who have contributed to this Bedford Way Paper show very clearly that this *can* be done. Situations in which the software dictates the learning programme leave teachers powerless and frustrated. Teachers must be allowed choices; they must decide how to use the power of the technology to achieve their own pedagogical objectives, taking into account their own classes and school situation. It is in these circumstances that we can look to the didactical variables which enhance computer use and learning for *all* pupils.

Throughout this volume there is a focus on education and *not* on the computer; a focus on task and process and *not* on performance. Both research and classroom practice are presented in an attempt to describe their complementarity and to achieve a balance between different ways of interpreting classroom events. As the SMILE teachers note in their contribution: 'working with pupils over a long period of time . . . provides teachers with insights which are not often apparent to an outside observer'. Their contribution and that of Sue Burns and Teresa Smart provide a wealth of insight into behaviour patterns and social interactions related to computer use and suggest teaching strategies that might help girls to develop their competence with computers. This is complemented by the account of the research findings from the Logo Maths Project (Sutherland and Hoyles) which describes in detail the different ways of interacting with the computer and differences in the nature of the collaboration between boys and girls. A contrasting approach to the question of pair work on the computer is described in the research reported here by Martin Hughes and his colleagues. Their quite surprising finding — that girls do better in mixed pairs than single sex pairs — is important since it forces us to take a more careful look at the question of single sex groupings and still more fundamentally at the nature and purposes of computer work in school. This then leads us to the last contribution in this volume by Richard Noss, who argues for a more pervasive 'cultural' approach to any consideration of computer use and computer performance and looks to a situation where there is a rich interaction between the software and children's intuitive mathematical understandings.

Finally, as editor of these papers, I would wish to acknowledge the danger that talking about gender-specific behaviour and male and female differences can give the impression of the existence of absolutes which then become self reinforcing. However, to make the distinction has allowed us to highlight differences in access to computers and sex stereotyping of the image of computers, as well as to begin to understand the social construction of computer use.

There is an urgent need for more research in this area, co-ordinated with teacher education and intervention programmes. With the computer revolution underway and the pace of change increasing, we have a duty to ensure that all children benefit from this change and that research informs the nature of the change.

Reference

Hofstadter, D. R. (1985), *Metamagical Themes.* New York: Bantam.

Review of the Literature

Celia Hoyles

Attitudes to computers

In a survey of the attitudes of 1,500 16 to 17 year-old pupils to computers and computing in Northern Ireland, Gardner et al. (1985) reported that the majority of boys and girls saw computing skills as essential for future prospects. However, significantly, more boys were of this opinion. In a survey of the attitudes of 1,600 4 to 16 year-old pupils, Wilder et al. (1985) found that boys and girls alike perceived the computer to be more appropriate for boys than for girls. In addition, although both sexes reported attitudes that were positive, more boys than girls liked the computer at all ages. Similarly, in a study in Scotland undertaken by Hughes and colleagues which predated the work reported in this volume, it was found that when 158 children of nursery and primary school age were asked about whether they thought boys would like computers more than would girls, the great majority associated computers with boys (Hughes et al., 1985). In follow up work with primary age children focusing on boy-girl interaction during Logo work, Siann and MacLeod observed that:

> Firstly the girls on the whole were less interested and motivated than the boys; secondly the girls were more disposed to turn to and seek help from the boys than the reverse; and finally, although the girls did seek help from the boys, they resented it when the help was given practically (i.e., by pressing the appropriate keys) rather than verbally. (Siann and MacLeod, 1986, p.137)

Chen (1986) after his survey amongst adolescents in the United States observed a certain ambivalence in the attitudes of females. Girls felt strongly that females were as competent as males but at the same time had more negative feelings about *their own* personal involvement with computers.

These and other findings (see, for example, Moore, 1986) suggest that intervention strategies which emphasize only gender equality issues will not necessarily translate into personal action for many females. Attention must be paid to the social influences affecting attitudes to computers.

The public image of computers

An important influence on attitudes to the computer is the image that the computer has within our society. Outside of schools there has been little research on male and female perception of computers. A survey carried out by the British Broadcasting Corporation in 1982 concerning the audience breakdown for two computer programmes showed an approximately 50 per cent split between the sexes. However, despite this equal division of the viewing audience between the sexes, it was discovered that 96 per cent of computer buyers were men. This suggests that men — and consequently boys — are likely to be the more active users of the technology in the home, so that computers tend to be seen as 'machines for men'. The type of software targeted for the home market reflects and reinforces this situation. It largely consists of games whose titles and themes exhibit gender bias; for example, they often concern war scenarios and physical adventures.

We must therefore ask what is the image of a successful computer user? In a study which explored the pictorial representation of males and females in popular computer magazines, Ware and Stuck (1985) reported that men appeared in illustrations almost twice as often as women; women were over-represented as clerical workers and sex objects, while men were over-represented as managers, experts and repair technicians. Only women were shown as computer-phobes. Of course to a certain extent these images give a true reflection of the present situation; for example Simons (1981) found that women are most frequently in low-grade jobs in the computing field and rarely in high positions. Women working in the field of computer science claim that they suffer from both implicit and explicit sexism which disadvantages them in terms of promotion (Lloyd and Newall, 1985). An outcome of this is that while careers in computing are very popular with boys, they are not considered by many girls (Culley, 1986). However, a recent study does suggest that the negative stereotyping of female computer scientists might be changing. Siann et al. (1986) investigated students' attitudes to, knowledge about and experience of computers. Half the respondents, randomly selected, received questionnaires describing a female computer scientist and the other half describing a male computer scientist. Aside from gender of the ratee, the questionnaires were identical. The researchers found that the target figure of a female computer scientist was rated more positively than a male computer scientist in eight out of 16 attributions. It will be of interest to see if this trend is revealed elsewhere.

Access to computers in schools

Throughout our society, parents are more likely to encourage boys than girls to use computers. Boys are much more likely to have a computer at home and have access to manuals and books. In a recent research project funded by the Equal Opportunities Commission (Culley, 1986) it was found that of the 974 fourth and fifth-year pupils who were asked about home computers, 383 (39 per cent) reported that there was a computer in their home. The gender difference in access to home computers, however, was very marked. Out of the 491 boys asked about home computers 261 (56 per cent) reported that there was a computer at home, while out of the 483 girls asked about home computers only 107 (22 per cent) had a computer at home. Differential access of boys and girls to home computers is therefore significant. This difference is also the case even amongst pupils taking computer studies options (ibid., p.34). The fact that girls taking computer studies examination courses are less likely than boys to have access to a home computer inevitably places them at a disadvantage.

Another aspect relating to access to computers is that boys are more likely to engage in computer use in clubs and extra curricular activities. Hess (1985), for example, reported that three times as many boys as girls were enrolled in summer computer programmes. In addition, the ratio of males to females increased with grade, cost and level of difficulty of the programme. A recent survey of 1,747 teenagers in the United Kingdom (Fife-Shaw et al., 1986) showed that boys used computers substantially more than did girls and that the sex difference was particularly strong for 'high-level' activities, such as programming. Because of their greater accessibility to computers, boys are able to build up a network of friends with whom to share ideas and experiences about computer use. Such networks confer approval and status on their members. Girls are rarely participants in these networks for several reasons: their lack of access to machines; because boys are reluctant to share their discoveries with girls; and because the computer groups frequently have an atmosphere and a language which are alienating for girls.

As well as considering experience with computers outside school hours, access to the hardware in school must also be taken into account. In a recent study in primary schools (Carmichael et al., 1986), it was noted that the boys took over the computers in free periods since girls did not want to compete with them. In class time, too, boys were more willing to volunteer to do things on the computer for the teacher. In one Grade 7 class it was reported that boys had complete control over the five computers and, as their confidence increased, that of the girls declined. At the end of the year the girls felt that they were so far behind they could never catch up and even commented that they never wanted a computer in their classroom again (Moore, 1986). In the EOC research project mentioned above it was found that through their greater familiarity and their general physical and social power, boys managed to secure for themselves a greater share of resources and teacher attention than did girls. Detailed observations of 42 lessons also revealed that in the discussion parts of lessons boys dominated, consistently asking more questions of the teacher and making more comments on the content of the lesson. Girls were also marginal to the class in a physical sense, often seated in groups at the back or sides of the room. Few teachers, it was noted, seemed to make any effort to counteract the tendency of boys to dominate lessons (Culley, 1986, ch.8). In the practical part of lessons boys would typically acquire the newest, most powerful computers — for example, those with disc drives and colour monitors — and girls would often be elbowed out of the way in the rush for the machines.

Use of computers in schools

In a useful summary of research in this area, Lockheed distinguishes three different computer applications: the computer as an object of study (programming, computer studies or computer literacy courses); the computer as recreation (game playing); and the computer as a general purpose tool (word processing, data base management, spreadsheets, graphics, music generation, and so forth) (Lockheed, 1985, p.119). She points out that males more than females use computers for programming and game playing but not for other computer applications. We must therefore examine gender issues within the context of these varied approaches to the educational uses of computers.

Firstly, optional courses of computer studies (where the computer is the object of study) tend to be dominated by boys. In 1985 in Great Britain for example almost three times as many boys as girls were entered for the Computer Studies Ordinary Level examination at age 16 years (43, 947 boys, 18, 538 girls). This proportion increased to almost five times as many boys as girls entered for the Computer Studies Advanced Level examination at age 18 years (7,670 boys, 1,652 girls). Moreover, at the undergraduate level the most recent figures for applications to British universities Computer

Science courses suggest that the situation is getting worse rather than better, with the proportion of female applicants dropping from 28.2 per cent in 1978 to 13.2 per cent in 1986; and the percentage of women actually enrolling similarly shrinking from 27.3 per cent in 1978 to 13.7 per cent in 1986 (UCCA, 1987). In fact there are fewer women applying for computer science courses now than nine years ago although the subject has doubled in size. The figures are illustrated in the table below. Similar situations are found in other countries (see, for example, Hawkins, 1984).

Table: UCCA Figures for Computing Science

Applications by year	*M*	*F*	*Total*	*% Female*
78	1677	659	2336	28.2
79	2470	918	3388	27.0
80	3331	1165	4496	25.9
81	3775	1287	5062	25.4
82	3855	1143	5331	21.4
83	4248	1083	5442	19.9
84	4450	992	5498	18.0
85	3872	626	4498	13.9
86	3486	534	4020	13.2

Acceptances by year	*M*	*F*	*Total*	*% Female*
78	1027	386	1413	27.3
79	1269	438	1707	25.6
80	1481	515	1996	25.8
81	1586	520	2106	24.6
82	1489	412	1901	21.6
83	1554	381	1935	19.6
84	1728	348	2070	16.8
85	1585	243	1828	13.2
86	1639	261	1900	13.7

Source: Universities Central Council on Admissions 1987

The image of these courses tends to be 'male' in both content and presentation and this could at least partly explain the data. The organization and teaching of computing in secondary schools is also a factor. This was investigated by the EOC project and in particular the reasons why male and female pupils chose to take or not to take computer studies options was explored. The project reported that computer studies was a very popular option choice for boys *and* girls but suggested that the *competition* for places in itself put girls off choosing the option. In addition, in deciding whether or not a particular pupil would be accepted on the computing course, the majority of schools took into account 'ability in mathematics' , even though it was generally recognized that this 'ability' was not necessary! The rationale was that it was an easy means of selection. This would seem to reinforce the strong link of computer studies in the minds of pupils with mathematics and science. In a great many schools computer studies is still in the hands of the mathematics department, but even as this becomes increasingly not the case the linkage still persists. Thus computers tend to be conceptually assimilated to the category of science, mathematics and technology and acquire some of the traditional qualities of differentiated interest amongst boys and girls. This is to some extent ironic since the professional use of computers has often very little to do with science and mathematics — although it must be said from the admittedly biased perspective of a mathematics educator a lot of the more interesting computer uses *are* to some extent mathematical!

There is now a move to treat the computer as a multipurpose tool in courses described variously as computer literacy or information technology awareness. These courses are, however, still *outside* the main school curriculum and, though possibly important for a transitional period (where there is a shortage of both distributed hardware and of teacher expertise), do not really address the fundamental question of *functional* computer use in schools (from the point of view of teachers and pupils). These courses can, moreover, create their own tensions: that is, tensions between time spent on 'regular' school work and computer work, and it is possible that such tensions might more strongly affect girls. Curriculum focused courses, in which the utility and power of the computer in learning a subject or range of subjects are displayed, seem to provide a more direct route to the greater participation of girls. This should not be taken to imply that the curriculum stays the same, rather the reverse. The point is that computer use should not be marginalized and separated from the other aspects of learning.

A crucial factor in considering computer use is mode of working. The use of the computer is often seen as isolating and devoid of interaction with others. We must allow flexible working arrangements which cater for peer collaboration as well as individual work. This then raises questions of effective working partnerships (see Hoyles and Sutherland, 1986 (a), (b); Hoyles and Sutherland, 1989, in press). It also raises questions as to what is valued by teachers. Carmichael et al. (1986) found that boys loved to work fast to complete the challenges set, but the girls lost interest as their efforts were not acknowledged. One girl commented 'Boys get commended for things that girls can do, but we complete it half an hour later, and it doesn't matter then' (quoted in Moore, 1986, p.7). Another factor is the sex of the teacher and the availability of female role models. Girls need to see females as competent, confident and enthusiastic computer users, although of course it must be recognized that the sex of the teacher is not a predictor of non-sexist practice!

The issue of single-sex computer classes also needs discussion. While hardware is restricted it might well be necessary to organize for girls to have uncontested access to the computer room. In their attitude survey in Northern Ireland, Gardner et al. (1985) found that girls in coeducational schools were more likely to be influenced by gender stereotypes in their attitudes to computers than were their counterparts in single-sex schools. However this conventional wisdom is questioned by Hughes and his colleagues in this volume, who suggest that 7 year-old girls do less well in Logo work only when working with another girl. This work is important in focusing on how group processes may influence individual learning paths. It might, however, say more about the acceptance of the task and the performance criteria of the task — the latter in this case were speed and number of key presses — than about gender differences.

References

Carmichael, H. W., Burnett, J. D., Higginson, W. C., Moore, B. G. and Pollard, P. J. (1985), *Computers, Children and Classrooms: a mulitisite evaluation of the creative use of microcomputers by elementary school children.* Ontario, Canada: Queen's Printer.

Chen, M. (1986), 'Gender and computers: the beneficial effects of experience on attitudes', *Journal of Educational Computing Research,* Vol.2, No.3.

Culley, L. A. (1986), *Gender Differences and Computing in Secondary Schools.* Loughborough: Department of Education, Loughborough University of Technology.

Fife-Shaw, C., Breakwell, G. M., Lee, T. and Spencer, J. (1986), 'Patterns of teenage computer usage', *Journal of Computer Assisted Learning,* Vol.2, pp.152-161.

Gardner, J. R., McEwen, A. and Curry, C. A. (1985), 'A sample survey of attitudes to computer studies', *Computers and Education,* Vol.10, No.2, pp.293-298.

Hawkins, J. (1984), *Computers and Girls: rethinking the issues.* Bank Street Technical Report No.24.

Hess, R. D. (1985), 'Gender differences in enrolment in computer camps and classes', *Sex Roles.* Vol.13, Nos.3/4.

Hoyles, C. and Sutherland, R. (1986a.), 'When 45 equals 60', *Proceedings of the Second Logo and Mathematics Education Conference.* London: Institute of Education, University of London, July.

_______ (1986b.), 'Peer interaction in a programming environment', *Proceedings of the Tenth Psychology of Mathematics Education Conference.* London: July.

_______ (1989), *Logo Mathematics in the Classroom.* London: Routledge (in press).

Hughes, M., MacLeod, H. and Potts, C. (1985), 'Using Logo with infant school children', *Educational Psychology.* Vol.5, pp.287-301.

Lloyds, A. and Newall, L. (1985), 'Women and computers', in N. Faulkner and E. Arnold (eds), *Smothered by Invention: technology in women's life.* London: Pluto Press.

Lockheed, M. E. (1985) 'Women, girls and computers: a first look at the evidence', *Sex Roles,* Vol.13, No.3/4.

Moore, B. (1986), *Equity in Education: gender issues in the use of computers. A review and bibliography.* Ontario, Canada: Review and Evaluation Bulletins, Ministry of Education, Vol.6, No.1.

Siann, G. and MacLeod, H. (1986), 'Computers and children of primary school age: issues and questions', *British Journal of Educational Technology,* Vol.2, No.17, May, pp.133-144.

Siann, G., Durndell, A., MacLeod, H. and Glissov, P. (1986), *Stereotyping in Relation to the Gender Gap in Participation in Computing.* Glasgow College of Technology.

Simons, G. L. (1981), *Women in Computers,* London: National Computing Centre.

Ware, M. C. and Stuck, M. F. (1985), 'Sex-role messages vis-à-vis microcomputer use: a look at the pictures', *Sex Roles,* Vol.13, Nos.3/4.

Wilder, G., Mackie, D. and Cooper, J. (1985), 'Gender and computers: two surveys of computer-related attitudes', *Sex Roles,* Vol.13, Nos.3/4.

My Mum Uses a Computer, Too!

Jeremy Burke, Julie-Ann Edwards, John Jeffries, Keith Jones, Judy Miln, Maggie Montgomery, Bridget Perkins, Andrew Seager and Helen Wright

About SMILE

The introduction into mathematics classrooms of pupil-centred and resource-based learning has essentially changed the traditional role of the teacher. The old 'chalk and talk' means of communication to passive, waiting children has been replaced by a style of learning where pupils are encouraged to accept responsibility for organizing their own work and mathematical thinking.

SMILE is an Inner London Education Authority mathematics project which encourages pupils' autonomy in their learning. At various times, the pupils will work as a whole class, in groups, in pairs or alone; this may be in response to structuring by the teacher, or as a result of decisions made by the pupils themselves. The teacher sets work individually for pupils, to draw out their confidence and abilities. The teacher keeps a record sheet for each pupil and can see at a glance which work has been attempted on various mathematical topics. The record sheet details individual pupil's progress, outlining areas where help and encouragement are needed. From time to time, teacher and pupil will discuss the pupil's progress and make decisions about future units of work. Flexibility is achieved by using a large and growing selection of over 1500 resource materials, including workcards, worksheets, posters, packs, audio tapes and microcomputer activities.

The project is controlled and developed by working groups of practising teachers. Resources introduced into SMILE classrooms are written, tried out and edited by SMILE teachers, based on the perceived needs of pupils in their classrooms. Current priorities include: bilingual pupils; girls and mathematics; and post-GCSE materials. Teachers are committed to providing genuinely equal learning opportunities and this is reflected in the continuing generation of new materials and the revision or rejection of existing materials.

SMILE's open-ended and investigative approach encourages a learning environment in which confidence in the learning process is more important than particular content. Pupils are expected to be fully in control of all the resources provided within a SMILE classroom, and to make decisions on the actual working styles that they choose to follow. The SMILE teacher's role is to work with the pupils to provide a rich, supportive and mathematically stimulating environment, within which pupils can develop their own confidence and understanding. This collaborative learning environment is found to be especially rewarding for girls. It encourages girls to put forward their own opinions for discussion and reflection, both to the teacher and to other pupils, working with them to complete a particular task. The discussion of work becomes an important part of learning. Aggression and competition are redirected in the positive challenging of each others's ideas.

Independence in controlling working time gives pupils experience in effectively managing their time in the lesson to the best possible advantage. It also allows them to determine the depth to which they will pursue a particular task. The result is that SMILE tasks are completed with confidence. The teacher does not act just as a provider of information, but often works as a co-learner in problem-solving situations. Pupils (and teachers) have the opportunity to set their own challenges.

The intrinsic flexibility of SMILE means that, in using its resources, no two schools across the country will have established the same approach towards learning mathematics. Rather, it is left to both pupils and teachers to establish the atmosphere within a SMILE classroom. Responding to the different personalities and tastes of pupils creates a secure environment where they are able to adopt a style of learning which is comfortable to them. In this situation, mathematical exploration need not be a frightening prospect, but a stimulating and rewarding experience.

It is in this context that work done by girls using computers in SMILE classrooms is to be seen. The computer is treated as just one other resource that the pupils may need to use to achieve an understanding of the task they are working on. The computer is integrated with other resources, and a corner will probably have been set aside for its use, with the screen positioned so that pupils can choose when to share their work with others, and when not. Research shows that work done by girls using a computer flourishes more readily within this more human context than in the typically formal atmosphere of a school computer laboratory; and, because the machine is situated in the mathematics classroom, the work done with it is seen as part of the mainstream mathematics curriculum.

Some classroom experiences

A busy SMILE classroom provides a wide variety of situations in which girls do mathematics confidently, with or without a computer. These will have been created as a result of a continued process of observation and reflection by the teacher. SMILE teachers are well used to reflecting on their classroom practice and on the learning undertaken by their pupils. Such reflection usually generates the need to make decisions which affect both the curriculum and the learning environment. Working with pupils over a long period of time under these conditions provides teachers with insights which are not often apparent to an outside observer. Behaviour patterns and social interactions, in particular, become evident. The accumulated experiences of SMILE teachers generate a range of appropriate materials and effective strategies which alter the classroom environment to make it a more favourable learning place for girls.

1. The SMILE classroom environment

The following accounts of SMILE classroom work, written by teachers and their pupils themselves, describe some of these. A collaborative teacher-pupil relationship is an essential feature of a SMILE mathematics classroom and is one of the reasons why these girls are able to feel successful in their work. A wide age and ability range from first to fifth year pupils (11 to 16 year olds) is represented. They describe working with computers in the classroom and how this has changed attitudes and learning situations for girls.

Simone

Simone was among nine girls in a mixed ability class of fifth-year pupils. She wanted to do well in mathematics but raced through each piece of work, often unconcerned if she did not understand it. Despite employing a variety of strategies to improve this situation, I felt I was having little effect. Even though Simone had a lively personality, she did not have close friends amongst the girls in the class, and she was the object of the boys' sexual innuendo and remarks.

I was using Logo in the classroom with three other classes and decided to make it available to this fifth-year class. Many of these pupils were completing coursework for their examinations. Simone, among others, had completed a vast amount of coursework and was interested in using the computer. At first she worked with another girl, Cheryl, but after two lessons, Cheryl returned to other work and Simone worked mainly alone. Having explored various commands and the wrapping facility, she began to draw a house in direct drive. She went on to draw various shapes in direct drive and, at this stage, I showed her how to write a procedure. In the following lesson she wrote her commands on paper as she drew her house again and used them for a procedure which she called 'HOUSE'. At the end of the lesson, we saved 'HOUSE' on disc.

In the next lesson Simone retrieved her procedure and was delighted to find she could draw the house again on the screen. She decided to add a path. This was also saved. During the following lessons an elaborate picture developed with a gate, fence, tree, flowers, cat and dog, each being a sub-procedure of one final procedure called 'SIMONE'. As she gained experience, Simone used circles and curves. By the end of the term the work was completed to her satisfaction.

Initially, Simone used Logo for seven consecutive lessons and then worked either on other work or Logo, depending on whether other pupils wanted to use the computer. We used two Commodore 64's in the classroom. Simone always set up the machine herself and loaded and saved her work.

Working on her own Logo task had two effects for Simone. Firstly, she found an activity at which she was able to persevere, overcoming problems to complete the goal she had set herself. The persistence she exhibited using Logo was in contrast to her approach towards her other mathematics work. She knew how she wanted each component of her picture to look and modified her procedures until she was satisfied with the outcome. She sometimes asked for help, but was deep in concentration for most of the time. This was a situation where she felt in control of her own learning, unlike her usual experiences of mathematics.

Secondly, Simone's role in the class and her self-image changed. In particular, the boys reacted to her in a different way and the content of their exchanges with her altered. They sometimes asked if they could sit with her to watch her Logo work. This she sometimes agreed to, but she always retained the position of power. Boys also asked for her help and advice when they were working on Logo themselves. Her knowledge and understanding of Logo were respected and with this respect came an increase in status accompanied by a considerable reduction in sexual innuendo from the boys. Within the classroom Simone is now more confident and self-assured.

Josie, Lorraine and Sarah

This class of fourth-year pupils is a group with whom I am confident about exploring new ideas. Computers are available for all their mathematics lessons. Any anxiety involved in doing mathematics seems to have disappeared for most of these students. I encourage them to work with one another as often as is possible, to investigate concrete situations when they do not understand abstract ideas, and to use Logo as freely as they can, given machine restrictions. When I worry that they may be spending too much time using the computer to do mathematics, I have to tell myself to relax and allow things to take their natural course as the work they do is always so good. I talk to them about the way I feel, but also tell them that my attitudes are probably old-fashioned. They agree!

I encourage this class to display their work themselves, and the walls of the classroom are a testimony to their rich mathematical imaginations. Three of these fourth-year pupils, Josie, Lorraine and Sarah, have been working together as a group since the beginning of the school year. Here they describe their work with Logo and their reactions to using a computer:

> We are doing a project called JOSIE where we first wrote the name across the screen, each letter separately, then altogether. Now we are writing it diagonally down the screen. We did make some mistakes (some letters were upside down or not in the correct place; we had lots of trouble with the E!) but after a few lessons working on it we finally got it all right. When we did all the letters we saved them, which we thought we could never do. If we couldn't save it we would spend all our time just loading the commands into the computer each time.
>
> Every day we have learned something new. It seems as though we have been doing it a long time, but we have really worked hard at it. Now we can have Josie's name on the screen at the touch of a button. We have it coming down the screen in a diagonal. We can print it in different colours or with different coloured backgrounds. We have really enjoyed it and are very pleased with what we have done.

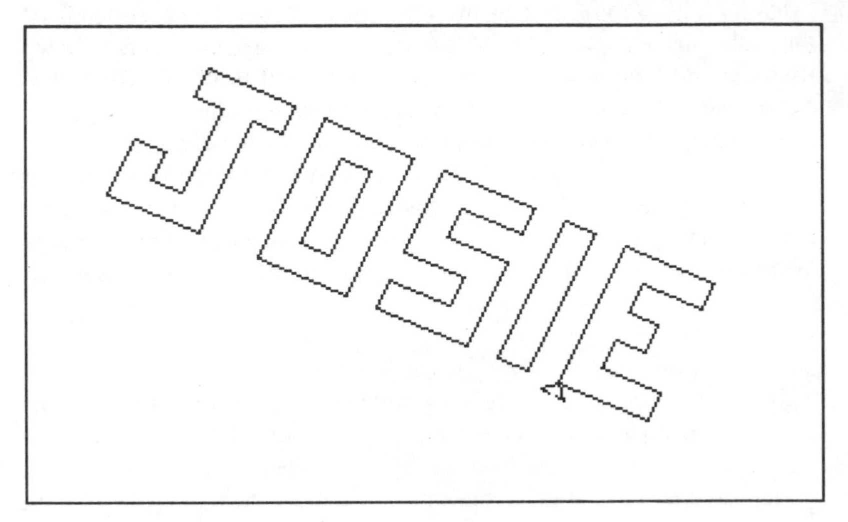

2. Screendump of Josie

When we started the work we thought it was a bit confusing but now we know what's going on. Doing Logo is very good; it makes a change from other maths. It makes it a little more adventurous and exciting and makes you really want to do maths. When you see a square on the screen, for example, you feel you have really achieved something but if you drew the same square on paper it would be dull.

Boys are always being encouraged to do special projects and trying new ideas in maths. Most girls just do the work that has been set for them and don't bother to explore new ideas. This is because they haven't enough encouragement. Our teacher praises us every time we do something different. With her encouragement, we could do anything.

Pretti and Sandra

This first-year class has had an Atari computer in the mathematics classroom for two out of three lessons a week over the past eight months. All pupils have had about two weeks' continuous lesson time using Logo in pairs. A rota system, initially devised, has now fallen into disuse and pupils tend to use Logo as part of their general mathematics work, either because they feel like it on a particular day or because there is a development from their work which needs Logo. At the outset, girls were the first to use the computers and, although the boys complained about this, they eventually came to terms with the situation.

Pretti is one of these first-year pupils. She is not very confident with her mathematics work and this may be due in part to English being her second language. She was working on a task which involved her in developing patterns by rotating a shape and drawing its outline at each stage. I suggested that she might like to use Logo to do this. Pretti seemed quite motivated by this idea and with another girl developed a procedure to draw an L on the screen. The following lesson, they were able to rotate the L about the bottom left-hand corner, and after some adjustment to their original procedure, about a point halfway up the left-hand side of the L. Eventually, confident in her ability, Pretti completed the activity with a tail recursive procedure that drew rotated Ls.

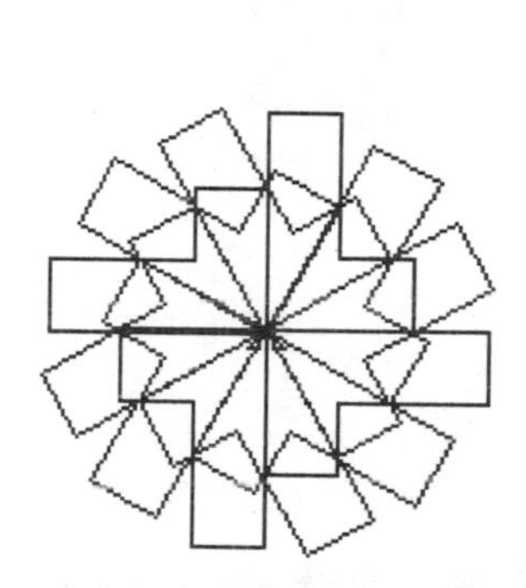

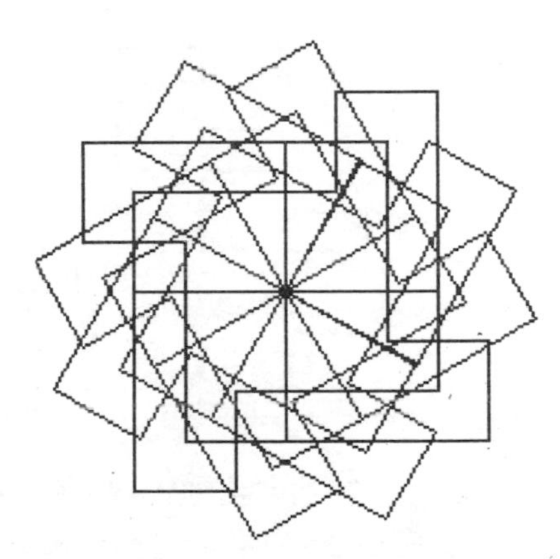

3. Screendumps of rotated Ls

Sandra, in the same class, seems to be almost addicted to Logo. She frequently hovers around the computer. She often acts as a 'Logo consultant' in the class and is respected in this. Her self-determined project at present is to develop pictures using circles and patterns from rotations of repeated designs. Currently, she is also working on a presentation of Logo to a first-year assembly. Sandra describes her reasons for enjoying Logo as follows:

> I enjoy Logo because it's easy. Anyone can do it. When you use Logo you are not tied down to a certain task or what the teacher tells you to do. When I am on the computer my mind is always ticking away, trying to do something new and adventurous. I have recently drawn a pair of spectacles and a snail and I have saved them on a disc. My Mum works with a computer at work and she enjoys it as well.

Spirals and mystic roses

I find that girls have a different response from boys to using a computer in the classroom. Usually, they need to have a reason for using it and are not very interested in the computer for its own sake. For this reason, our most successful use of computers has been in the extension of other work.

One experience of this was when a group of girls used Logo to draw a variety of spirals after investigating spirals with pencil and paper.

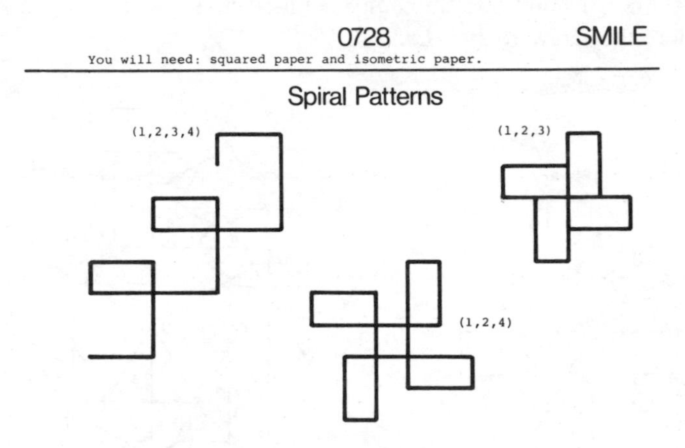

4. Spiral patterns (cover of SMILE booklet)

They became very involved in the work, finding out how to use variables in procedures effectively, and how to draw different arms of their spirals in different colours. Most of their final program was created by trial and error and, unwittingly, they were working at a much higher mathematical level than they normally do.

Another occasion when the computer proved to be an invaluable aid to mathematical understanding was when a group of girls had been investigating mystic rose shapes using the MicroSMILE program ROSE. They had generalized a relationship between the number of lines and dots on the shapes quite quickly, but they continued to use the program. When I questioned them about this, they said, 'Watching it makes you see why the formula works in two different ways'. The computer had allowed them insights into their generalization.

My experience has been that girls prefer to use a computer to follow up some work which they have already started. I have to make a considerable

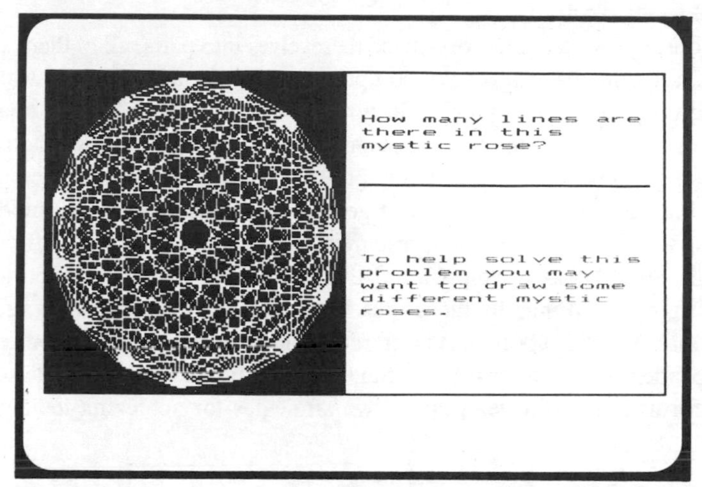

The problem is set . . .

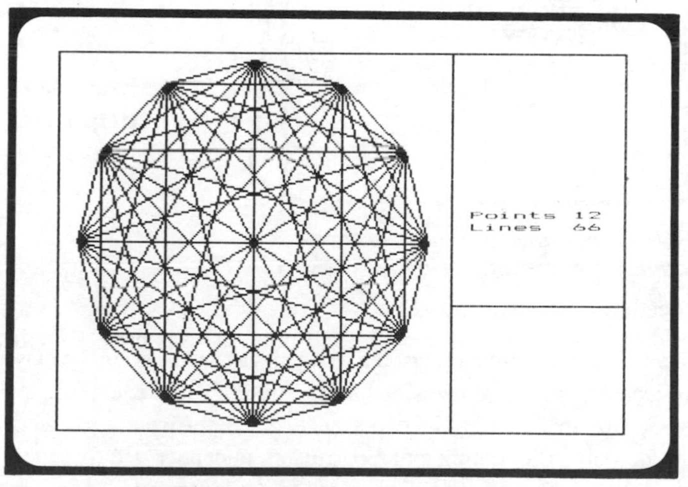

. . . the user has drawn a simpler rose.

5. Mystic rose patterns (from MicroSMILE booklet)

effort to create the time and space in the classroom for girls to use a computer when there is a purpose for them to do so. The results are well worth the effort.

Nicola and Andrea

My fourth-year class have organized themselves into pairs, all of them single sex, and pupils have agreed to co-operate on a long-term basis in working at the computer. Each pair is, in turn, given a block of lessons totalling four or five hours, on a rota system, to work at something of their own choosing. Some pairs exhaustively pre-plan their work whilst others work within a series of loosely defined goals that are developed and modified during the sessions. The main resources available for resolving difficulties are the SMILE Logo Help Cards[1] and the group's collective experience. By keeping to a minimum those situations in which the teacher must provide information, the approach is in keeping with the encouragement of independent learning in SMILE. Students seek the most appropriate source of information and develop their own strategies for achieving their goals.

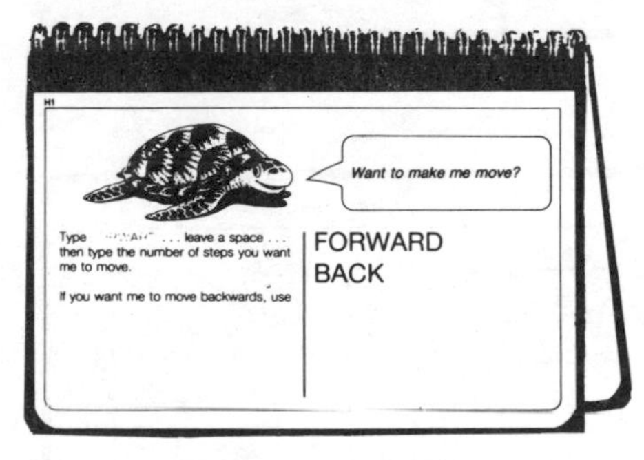

6. Example of Help Cards

The Help Cards provide immediate access to information pupils require in their work with Logo.

Nicola and Andrea were amongst the first pairs to use Logo in the classroom. This reflected a deliberate decision on my part to try to create positive female role models by investing initial experience and expertise in girls who would then be perceived as competent Logo users and peer group experts.

I was able to observe them closely at the computer and I was impressed with their way of working, their interest in exploring new commands and the resolve which they brought to the subsequent activity of making their own sense of these commands.

1 The Logo Help Cards and posters are part of a pack of materials *Mathematics and Logo,* produced by teachers through the SMILE Centre, Middle Row School, Kensal Road, London W10 5DB.

At the beginning of one session they asked if they could do something other than drawing. I was unable to answer them at the time as I was involved in helping another pupil. When I arrived at the computer, the girls had taken Logo arithmetic into their own hands and typed in '2+3'. They were looking at an unhelpful error message which read 'You don't say what to do with 5'. I made two suggestions which I thought would give them some ideas on Logo arithmetic.

Firstly, I explained how the PRINT command worked and drew their attention to a poster from the SMILE Logo Pack that sets the challenge of printing the numbers from 1 to 10 using only the digits 2 and 3 and the four arithmetic operations.

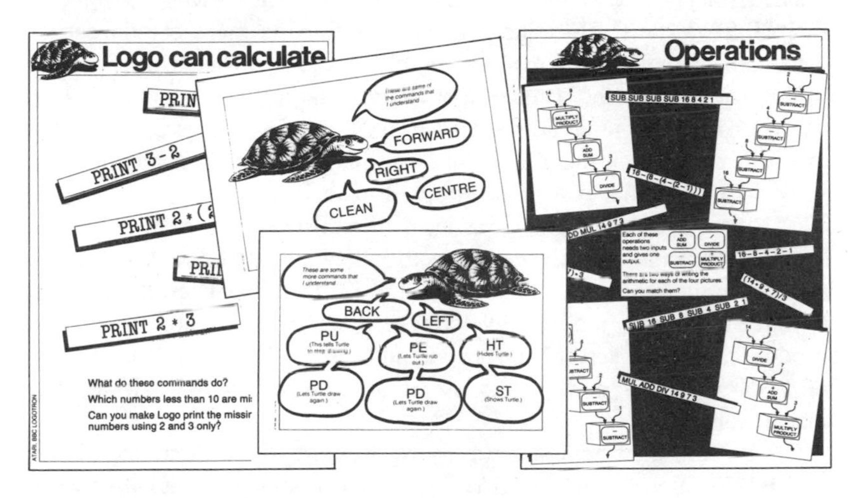

7. Logo poster

Secondly, I showed them a procedure which usefully employs arithmetic:

```
TO L :LENGTH
FD :LENGTH
RT 90
FD 2* :LENGTH
END
```

As Nicola and Andrea had not previously met the idea of a variable, this procedure was difficult for them, but it did provide a challenge in a genuine context for arithmetic. So they spent the rest of the two-hour session making sense of the ideas in L. They worked almost entirely interactively, trying out and then modifying their work until they were able to control this new aspect of Logo.

Firstly, they defined a procedure which they called PEN, a simplification of L.

```
TO PEN :LENGTH
RT 90
FD :LENGTH
END
```

Then, they replaced FD 2* :LENGTH in L by the PRINT statement, PRINT 2* 3, thus relocating arithmetic within its more certain earlier environment. After more exploration they got to know the syntax and the difference between the name of a procedure and a variable; why PEN on its own without an input fails and what PEN 50, PEN 100, etc., draw. This involved me helping them to act out, by 'playing Turtle', exactly how the screen Turtle evaluates a procedure with a variable input.

PEN gave them the idea of drawing different sized squares by absorbing the procedure into the REPEAT structure they already knew. Once again, they found themselves having to make sense of this; they knew they wanted to make the Turtle 'repeat PEN 100 four times' but were not sure how to produce this on the screen. Having only just come to terms with the working of PEN 100, they now had to use it as an instruction inside another instruction!

Once again, their approach was to 'try and see' — amongst other things REPEAT 4 PEN 100 and REPEAT 4 [PEN]. Everything fell into place when they suddenly realized that PEN 100 was in some ways an instruction just like FD 50 or RT 90. 'Oh, REPEAT 4 [PEN 100], is that all?' Nicola suddenly said. She tried it, saw that it worked and added delightedly, 'Oh, it's so easy'. They now typed in sequence:

```
REPEAT 4 [PEN 200]
REPEAT 4 [PEN 300]
REPEAT 4 [PEN 400]
```

and they were very pleased to see that, as they predicted, this drew a series of squares.

Gaining in confidence, Nicola suggested they do the pattern the other way round and they used their newly found skills to define PENL.

```
TO PENL :LENGTH
LT 90
FD :LENGTH
END
```

At the end of the session, both girls took away screen dumps and were tremendously pleased with their achievements. Perhaps, most importantly, they gained some insight into the considerable satisfaction that is derived from solving one's own mathematical problems.

Louise and Stacey

This fourth-year class had some experience in their first three years using Logo on computers in a computer room. Throughout this year there has been an Atari available for two out of four lessons each week. At the beginning of the year there was a general antipathy towards using Logo but an amount of enthusiasm for using MicroSMILE programs. I think this was a product of their previous use of computers. As a means of encouraging a more adventurous use of the computer, I suggested to pupils that they might like to extend their mathematics work using Logo, e.g., to do a spirals investigation which they had all enjoyed.

Louise and Stacey are pupils of about average ability in this class who had achieved relatively little in their third-year. Earlier this year, they decided to do a project on probability. From this developed the question 'Do Tarot cards predict the future?' They looked at the probability of achieving different arrangements of cards. However, fractions were proving a difficulty in working out and understanding the probability of combined events. After a long period of looking at equivalent fractions and some discussion on the multiplication of fractions, I suggested that they might like to write a Logo program to work out their fractions. This led to two weeks' work developing a program that multiplies two fractions. The girls tell their own story:

> In a magazine we saw a page containing different problems and challenges. We were quite interested in one problem that asked us to throw a dice ten times and see which number arose most frequently. In our maths lesson we got a dice from the cupboard and threw it to each other. Our teacher suggested keeping records. We groaned a bit because it involved putting pen to paper but we found it all very interesting. We told our teacher that we liked probability and he agreed with us, which we were glad about.
>
> Our probability project progressed more and more until it involved fractions.

Our first reaction was to give up but our teacher suggested using Logo. Firstly, we began experimenting. We learned a lot from mistakes and programs that went a bit wrong. Instead of asking our teacher to help us out, we tried and tried until it looked right to us. He never tried to interfere or tell us that we had crazy ideas.

Suddenly, we had a brainstorm. We made a big program called ADDFRACT, containing lots of smaller ones — seven of them. ADDFRACT is a program where you can add and multiply fractions. It's an easy and fun way of learning.

We think using Logo is a great idea because if we ever forget how to work out fractions all we have to do is get our program and it won't shout at us. We hope to do lots, lots more on Logo as it is a really enjoyable way of learning any kind of maths.

Evanthia and Sally

This year, I have two computers in the classroom with my third-year mathematics class. In previous years with other classes, I had insisted that pupils worked on the computer in stable pairs and on a rota system giving pupils equal access. This helped me to become confident in using computers in the classroom and, I thought, reduced the possibility of arguments amongst pupils over their use.

It soon became apparent that this third-year class, together with their previous mathematics teacher, had already developed a much more sophisticated system of their own. Pairings had become more flexible, particularly amongst the girls; the system for access involved considerable negotiation between pupils if a particular pair wanted to continue a Logo project. Though the girls were in the minority in the class, it became clear that it was through their influence that this sophisticated structure had arisen.

As the year progressed, I became more interested in the Logo work of four of the girls because of the nature and degree of collaboration involved in their work. Sally and Sarah had been deeply involved for a long period of time in developing a LOGO fish tank full of exotic fish and aquatic plants. Evanthia and Marina began with a fashion theme, using Logo procedures for various 'tops' and 'bottoms' which could be swapped about to produce a wide variety of different outfits.

During the year, a classroom friendship developed between Evanthia and Sally which superseded those with Sarah and Marina and they decided to work together in their Logo work. Each was committed to their individual objective and so, instead of embarking on a mutual beginning, they worked jointly on each other's work. I was intrigued by the special sort of negotiation and collaboration this required. Neither girl lost her resolve over her original goal, yet their individual projects clearly benefited from the mutual sharing of ideas while the girls worked together. There was an equal share of time spent on each of the projects, indeed equal enjoyment of each other's work.

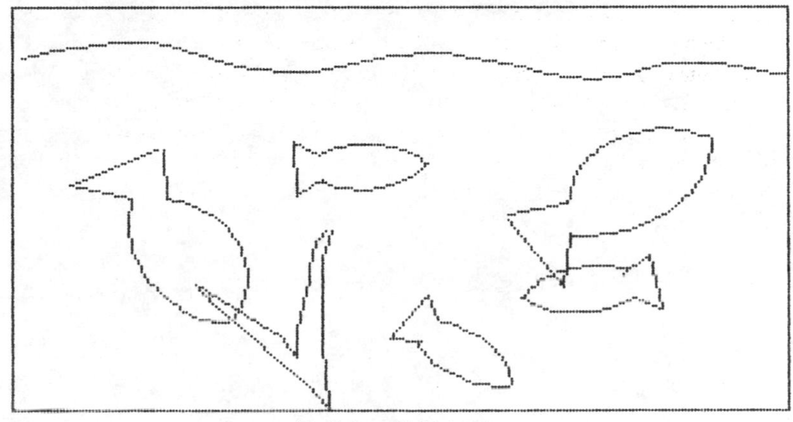

8. Logo fish tank

I found myself realizing that it is the nature of working together which is of greater importance to girls than having a mutual or common goal.

Newtiles

Two first-year girls came to my classroom after school one day, intent on completing some SMILE work. I was exploring the MicroSMILE Newtiles program and they were intrigued with what I was doing. I gave them a quick explanation about designing the tile and building up a floor pattern using rotating and reflecting tiles. They both wanted to use the program and initially worked together on their first attempt. Their tastes as to what made a good tile or pattern differed considerably, however, so they decided to work together on each other's choice of tile. They talked a great deal about the work, sharing both technical and aesthetic ideas on the patterns. They also asked my advice on a few occasions, which they sometimes acted upon or chose to ignore.

By the end of two hours' uninterrupted work, both girls had each completed a pattern to their own satisfaction and saved it on disc. In their mathematics lesson they worked together again to produce more tiled patterns. The main attraction seemed to be that there was not an immediately desirable end product, simply an opportunity to explore the program until they achieved their own satisfying result, and to discuss their work with each other as they went along.

9. MicroSMILE Newtiles Program

Conclusion

The flexibility inherent in classrooms where teachers use SMILE to teach mathematics is evident in the accounts above. It gives both teachers and pupils a variety of choices in determining exactly how and when computers are used. Each description is school-specific, indeed classroom-specific, and the personalities of pupils and teachers provide a different perspective in each case.

Common to all, though, is a determination on the part of teachers and their pupils actively to work against traditional sexual roles with respect to the computer. Women teachers use their own reactions to computer use to create positive role models themselves; Sandra finds a positive role model in her mother; male teachers encourage girls to be the first in the class to use computers; the boys in Simone's class changed their attitude towards her; and Josie, Lorraine and Sarah tell us emphatically that girls are generally not given enough encouragement to take up challenging ideas.

9. MicroSMILE Newtiles Program

Central in creating an environment supportive to girls using computers is the pupil-centred nature of the curriculum. Allowing pupils autonomy in their learning gives them control and therefore power over that learning. Computers, incorporated as resources for learning under these conditions, are also subject to this control. Competence and confidence in using resources become a matter of course for the girls in these classrooms.

Furthermore, pupils in SMILE classrooms are able to work on mathematics developed from their personal interests. This encourages girls to use a style of learning mathematics, both at and away from the computer, which is personalized and under their own control. Sally, Sarah, Evanthia and Marina were able to organize the computers and their time on them so that all four girls collaborated initially on the two projects. Confidence in doing this generates confidence in doing externally imposed tasks.

Peer collaboration, and collaboration with the teacher as a co-learner, is another product of a more pupil-centred curriculum. The classroom accounts seem to indicate that this collaboration is crucial to girls' effective mathematics learning. There is no evidence whatsoever of competitive learning. Yet highly individualistic goals are attained as readily as joint goals, in the cases of Evanthia and Sally and the two girls using the Newtiles program. Clearly, individuality and independent aims are not solely a product of competitive activity.

Traditionally highly-rated computer activities are absent from the descriptions of this classroom work. The tasks undertaken by the girls are not complex graphical displays. On the contrary, they are tasks borne out of the girls' own creative interests — Simone drew an elaborate picture — or out of a need for help — Louise and Stacey needed a program to help them multiply fractions.

What is described, though, is a depth of understanding of the task undertaken, a growth in confidence as a result of this, the feelings of excitement and joy of success. The term 'adventurous' is used: it is one we often leave out of our images of girls using computers. Yet these girls use it to describe their own work and feelings about the mathematics they do on a computer!

In summary, there are several strategies evident that SMILE teachers use to create a situation in the classroom where girls feel confident and supported in their use of the computer. They are:

- making the mathematics curriculum pupil-centred
- providing the opportunity for collaborative work on joint or independent activities
- giving credence to different styles of learning in the classroom
- using female role models and positive images of girls using computers
- providing situations where girls are able to extend their own work to a computer activity
- valuing the mathematics learning involved in computer activity rather than its visual complexity
- allowing girls to organize the resources to suit their own styles of learning.

Girls, Boys and Turtles: gender effects in young children learning with Logo

Martin Hughes, Ann Brackenridge, Alan Bibby and Pam Greenhough

The aim of the project described in this chapter[1] *is to examine the way in which the gender composition of pairs of young children influences the way they interact with the Logo turtle robot. The project involved 30 pairs of seven-year-old children: 10 boy-boy pairs (BB), 10 girl-girl pairs (GG) and 10 mixed pairs (BG). The initial findings showed no difference between the performance of children in the BB and BG groups on a standard task; however, the peformance of girls in the GG group was significantly below that of the other two groups, both when the girls performed as a pair and when they subsequently performed as individuals. Various explanations for the effect are discussed, with the present position being that an adequate single explanation has not yet been identified.*

The issue of gender and computing first came to our attention a few years ago when we were carrying out a study in which a simplified version of Logo was introduced to a group of 15 six-year-old children in a severely deprived area of Edinburgh (Hughes, et al., 1985; Hughes, 1986). In this study the Logo sessions were characterized by high levels of concentration, collaborative problem solving and mathematical discussion, and the children made significant gains over a period of five months on standard tests of number and shape. However, these significant gains were found only for the boys, and not for the (admittedly very few) girls in the study, suggesting that the Logo experience had been less valuable for the girls than for the boys.

1 It was presented as a paper at the Second Conference of the European Association for Research on Learning and Instruction held in Tubingen, West Germany, in September 1987.

The issue of gender also arose in another study we carried out in Edinburgh, in which we interviewed over a hundred children from different social backgrounds on their ideas, experiences and attitudes concerning computers. We found that by seven years of age, boys were already much more likely than girls to have a computer at home and to use it regularly (Hughes et al., 1987). We also found substantial sex-stereotyping at this age, with a large proportion of seven-year-olds claiming that boys are more likely than girls to 'like computers' and 'be good with computers'. Clearly, the issue of gender and computing can be traced back to the very start of compulsory schooling, and maybe even further back still.

The educational problem of gender and computing is related to the more familiar problem of girls being less interested than boys in subjects such as mathematics, science and technology, itself a serious cause for concern (Whyte, 1986). At the same time, the problem with computing is potentially far more pervasive. One of the more interesting educational developments of recent years lies in the way computers are being used across the curriculum, often making radical alterations to the way traditional subjects are being taught. If girls are in general less familiar with computers than boys are, or if children strongly identify computers with boys rather than girls, then the increasing use of computers across the school curriculum will have profound effects on the education of girls.

One solution which is frequently put forward whenever boys appear to dominate a particular part of the curriculum is that girls should be separated from boys and encouraged to work in single-sex groups. This proposal sounds reasonable, although there is in fact little hard evidence to support it: indeed, with the exception of the present study, there have been virtually no systematic comparisons of what actually happens when children learn with computers in single-sex and mixed-sex groups.

While our concern in this study is therefore with the specific issue of the gender composition of groups, this is part of a wider interest in the area of social interaction at the computer keyboard. Many of the earliest experiments in computer-aided learning reported in the 1960s (e.g. Suppes and Morningstar, 1969) took place in university laboratories where there were sufficient resources for each child to have their own individual computer terminal. Thus, when micro-computers first started to arrive in schools in the 1980s, it was frequently assumed that children would usually be working individually: indeed many educationalists still consider that one of the most important contributions that computers can make to education is in the vastly increased opportunity they provide to tailor instruction to the needs and ability level of individual children. However, the relatively small number of computers actually available in schools has meant that most children work at computers in small groups. For example, a recent survey of primary schools in the English county of Hertfordshire (Jackson et al., 1986) found that most children were using computers in groups of two, three or four, although, interestingly enough, this was related to the type of software in use: children were much more likely to work individually when using 'drill and practice' software. Indeed, many teachers now see positive benefits for children from working in small groups at the keyboard, pointing in particular to the way in which certain types of software (such as adventure games and simulations) can generate large amounts of animated discussion and collaborative problem-solving amongst small groups of children — an end frequently seen as desirable in itself.

There is thus a small but growing body of research on children's social interaction at the computer keyboard. In some studies, the focus is on whether working in groups actually does lead to improved individual performance (e.g. Fletcher, 1985; Light et al., 1986); this is, in fact, an old research problem which is being revisited via the microcomputer. Other studies are starting to look at the influence of factors such as the size, ability composition and gender composition of the group on group performance and interaction (e.g. Potter, 1985): the present chapter falls into this second category.

The present study

Our study thus follows on directly from the Edinburgh study we have described above, in which six-year-old children learnt to control the turtle robot by using a simplified version of Logo. We wanted to follow up in a systematic fashion the Edinburgh finding that girls were less affected than boys by this early introduction to Logo. More specifically, we chose to do this by looking at whether the gender composition of pairs of children affected their performance, i.e. we wanted to know whether there were any systematic differences between boy-boy, boy-girl, and girl-girl pairs.

The design which we adopted to answer this question differed quite markedly from that used in many other Logo studies (including our own previous work). In order to optimize the Logo learning environment, such studies have tended to use a relatively small number of subjects, over a long period of time, performing problems which they have selected for themselves. Because our focus was on making systematic comparisons between groups, we wanted sufficient numbers of children to allow statistical comparisons to be made (and hence a minimum of 60 children); and we wanted to study their performance on a standard task which we had chosen for them. The relatively large number of children involved, together with our desire to record in detail every aspect of the learning process, meant that we could look at the children only over a short period of time (less than an hour altogether). Clearly, these constraints on the study may not appeal to other Logo researchers, in that it may not fit their ideas of what Logo is and how it should be studied; equally clearly, such constraints are necessary if we want to obtain more systematic knowledge about certain aspects of the Logo learning process.

More specifically, the 60 children were divided into three matched groups of pairs so that ten pairs were boys-only (BB); ten pairs were girls-only (GG) and ten were mixed (BG). The groups were matched overall for age and ability, with the two children in each pair being of similar ability; ability for these purposes was based on their teachers' rankings. The children spent three sessions of 15-20 minutes working with the turtle. During the first session each pair familiarized itself with the turtle in a totally unstructured situation: the children were shown how to make the turtle move backwards and forwards, left and right, and then allowed to move it as they wished over a large board. In the second session each pair was asked to carry out a specific task, that of steering the turtle around an obstacle course mounted on the board. In the third session, the children again attempted the same obstacle course, but this time they were working individually. Each session was introduced by a male researcher who stayed in the room with the children but made minimal subsequent interventions, and was apparently occupied with some other work. The sessions with the turtle took place in a quiet area away from the classroom, and were separated by about a week. All sessions were videotaped, and each keypress on the computer was automatically recorded on a disk-based dribble file.

The sessions with the turtle were preceded and followed by a semi-structured interview with a female interviewer. In the first interview they were asked about their previous experience with computers (and related objects such as robots and remote controlled cars); their own attitude towards computers, and whether they associated computers with boys or girls. They were also tested for their understanding of left and right. In the second interview the questions concerning the children's attitudes and sex stereotyping were repeated, as was the assessment of left and right. In addition they were asked various questions about their sessions with the turtle.

The main findings

The main findings of the study were both clearcut and unexpected.

Table 1 shows how the three different groups (BB, BG and GG) performed in Session 2, when they were required to carry out the obstacle course task in pairs. The top row of the table shows simply whether the pair was able to complete the task or not in the 15-minute session; as can be seen, virtually all the BB and BG pairs were successful, compared with only two of the ten GG pairs The middle row of Table 1 shows a more fine-grained measure of how well each pair had done on the task, this measure being based on either their time to completion or their position on the course when the session finished, with a maximum high-score of 7. As before, the performance of the GG pairs was well below that of the BB and BG pairs. The final measure shown in Table 1 is the mean number of times that the turtle crashed into an obstacle on its way round the course. Again the pattern is the same, with the GG groups crashing the turtle nearly twice as often as the other two groups. It should be pointed out that the children did not appear to be crashing the turtle deliberately, and turtle crashes were usually accompanied by groans or other expressions of dismay! For all three measures shown here, the differences between the GG group and the other two groups are highly significant statistically; there are no significant differences between the BG and the BB groups.

Table 1: Pair performance on task (Session 2)

	BB *n=10*	*BG* *n=10*	*GG* *n=10*
Number of pairs to complete	10	9	2
Means 'finish' score (max=7)	5.1	5.2	2.5
Mean number of crashes	7.2	8.7	16.5

GG v other p<0.001 except for GG v GB crashes where p=0.01

Table 2 shows the children's performance on Session 3, when they were required to complete the task as individuals. Table 2 contains an additional column compared with Table 1, as the performance of the boys in the mixed group — B(G) — is shown separately from that of the girls in the mixed group — G(B). There was no time limit in this session, and the children were allowed as long as they wanted to complete the task: as a result it is possible to show the mean overall time for each group to complete the task, the mean number of discrete turtle moves to complete the task, and (as before) the mean number of turtle crashes on the way round.

Table 2: Individual performance on obstacle course task (Session 3)

	BB *n=20*	*B(G)* *n=10*	*G(B)* *n=10*	*GG* *n=20*
Mean time to complete	9.0	7.7	9.3	15.4
Mean number of turtle moves	59.1	52.9	58.7	85.9
Mean number of crashes	6.6	4.5	5.4	15.2
Mean 'finish' score (max=7)	5.5	6.0	5.4	3.7

GG v others $p<0.001$ All other differences N.S.

As can be seen, the pattern of Table 2 is almost identical to that of Table 1, with the performance of girls in the GG group being significantly worse than that of the other three groups. It is worth noting that the individual performance of the girls who were in mixed pairs is significantly better than that of the girls who were in girl-girl pairs, but is not significantly different from that of the boys.

Discussion

The findings reported above are quite unexpected, in that nothing similar appears to have been reported previously in the research literature. Indeed, it is often suggested that gender effects are unlikely to be found in early childhood, and that gender only becomes a serious issue during the secondary stage of schooling (11 years and over). The present findings show instead that strong gender effects can be found with children as young as six or seven years, thus raising the possibility that they may be present at an even earlier age.

The findings are also potentially controversial, suggesting as they do that girls need boys more than boys need girls, and bearing directly on the important classroom issue of single-sex grouping as a means of overcoming the disadvantage of girls in maths, science and technology. It is often assumed that girls will perform best in such subjects if they are removed from the supposedly inhibiting effects of boys and allowed to work in single-sex groups. Our findings, however, suggest that in fact the girls who worked in single-sex groups were at a serious disadvantage compared both with boys and with girls who worked with boys. At the same time, of course, we must be very tentative in assuming that the findings of a single study, which took place outside the classroom over a short period of time with only 30 pairs of children, can be readily generalized to a wide range of differing classroom situations. Further research is clearly needed to discover just how far the present findings can be generalized.

While the present findings are relatively clearcut, what is not so clear is the most appropriate explanation for them. However, two main types of explanation can be distinguished.

The first kind of explanation which should be considered is that the findings reflect a fundamental (and possibly innate) *cognitive* difference between the sexes, such as that boys have greater spatial awareness than girls. There will be psychologists and educationalists who will be only too ready to interpret our findings in this way. There are, however, two points that must be made in response to such an explanation.

The first point is that there is no evidence in this study for any initial cognitive difference between the various groups. True, we did not collect many pre-test measures of the children's cognitive functioning, not least because it was by no means apparent what measures would be most appropriate. However, in one relevant area, that of left/right understanding, detailed testing was carried out. Knowledge of left and right was tested with regard to the child's own orientation, that of the interviewer seated opposite, and from the point of view of inanimate objects placed between them. In no cases were any significant differences found between the groups.

The second point to make is that a purely cognitive explanation is not sufficient by itself to account for the particular findings of this study, in which the performance of the girls was directly affected by the gender of their partner. In other words, a cognitive explanation must be augmented by some kind of explanation based on the *social interaction* within the groups. For example within a BG pair passive observation of a superior (boy) model or tutoring by a more able (boy) child could account for the pattern of findings. In these cases the superiority of the boys would, however, need to be established, as well as evidence for the proposed teaching effect.

The second main type of explanation to be considered is that the findings reflect *attitudinal* differences between the sexes to the situation. It could be argued that in our society technology in general and computers in particular have a strong gender bias towards males, and that children learn from an early age to associate computers with boys and men: more specifically it could be argued that the task used here (that of steering a remote-controlled wheeled vehicle round a track) is more of a 'boy's task' than a 'girl's task'. The present findings would therefore be explained in terms of unfavourable attitudes to the task arising from a lack of experience, interest, motivation or confidence on the part of the girls. Indeed, it might even be argued that in describing the performance of the GG pairs as 'inferior' to that of the other groups we are making unjustified assumptions as to the value of such computer-based activities.

As before, two points need to be made in response to this type of explanation. The first is that while it has been found elsewhere that boys have greater access to home computers than do girls (see, for example, Hughes, et al., 1987), this was not true of this particular sample. Furthermore, our initial interviews showed the boys and the girls to be equally enthusiastic about computers before the experiment began. In answer to the question 'Do you like computers?' only one child, a boy, answered 'No'. Similarly girls were equally confident about their own personal abilities with computers. There was, however, some suggestion of an attitudinal difference between the boys and the girls in answer to a question about which sex in general was better with computers. While most of the children answered that there was no difference between boys and girls, the boys were more likely to show a positive preference for their own sex: this result, however, fell short of significance at the 0.05 level.

A similar picture emerged from our final interviews, where there was virtually unanimous agreement from both boys and girls that the sessions had been extremely enjoyable. Boys and girls also judged their own performance in equally favourable terms. There is, therefore, little evidence from the interviews for the assumption that strong attitudinal factors influenced the performance of the children. However, there may exist a more subtle difference in confidence which would become apparent from an exhaustive analysis of the behaviour exhibited during the sessions.

The second point is that, even if we had found that girls were in general less interested in or experienced with computers, such differences by themselves would not be able to account for the particular pattern of findings reported here, in which the performance of the girls was affected by the gender of their partner. As with the purely cognitive type of explanation discussed above, an explanation primarily in terms of attitudes would need to be augmented by some hypothesis concerning the *social processes* at work in the groups. For example, the fact of being in a GG pair working on a task perceived as 'male' might result in the GG pairs feeling less able to engage with the task or partake in the range of interpersonal exchanges which characterize the more valuable learning sessions: these exchanges would, however, need to be identified and evidence sought for diminished involvement in such exchanges by the GG pairs.

Clearly we shall be in a much better position to provide an explanation for our findings when we have completed our detailed analyses of the videotapes, and the results will be reported as soon as they are available.

References

Fletcher, B. C. (1985), 'Group and individual learning of junior school children on a microcomputer-based task: social or cognitive facilitation?', *Educational Review,* Vol.37, No.3, pp.251-261.

Hughes, M. (1986), *Children and Number.* Oxford: Basil Blackwell.

Hughes, M., Brackenridge, A. and MacLeod, H. (1987), 'Children's ideas about computers' in C. Crook and J. Rutkowska (eds), *The Child and the Computer,* London: Wiley.

Hughes, M., MacLeod, H. and Potts, C. (1985), 'Using Logo with infant school children', *Educational Psychology,* Vol.5, pp.287-301.

Jackson, A., Fletcher, B. C. and Messer, D. (1986), 'A survey of microcomputer use and provision in primary schools', *Journal of Computer Assisted Learning,* Vol.2, pp.45-55.

Light, P., Foot, T., Colbourn, C. and McClelland, I. (1987) 'Collaborative interactions at the microcomputer keyboard', *Educational Psychology,* Vol.7, pp.13-21.

Potter, F. (1985), 'Factors affecting the quantity and quality of group discussion with language and reading microcomputer programs', Paper presented at the British Educational Research Association Conference, Sheffield, 1985.

Suppes, P. and Morningstar, M. (1969), 'Computer-assisted instruction', *Science,* Vol.166, pp.343-350.

Whyte, J. (1986), *Girls into Science and Technology.* London: Routledge and Kegan Paul.

Gender Perspectives on Logo Programming in the Mathematics Curriculum

Rosamund Sutherland and Celia Hoyles

It has been suggested that the computer should as far as possible be integrated into the school curriculum, both to enable the present curriculum to be taught and learned more effectively and as a way of changing the curriculum in the light of the availability of the new technology. In mathematics classrooms a range of software is being used — from content specific special purpose software to more general applications software. In this chapter we shall consider the gender implications of introducing programming as part of the mathematics curriculum. We shall argue as mathematics educators that programming can have an educationally beneficial influence on the nature and content of school mathematics and this can be to the advantage of girls. As part of the discussion we shall look at the meaning of programming, different styles of programming and different modes of working at the computer.

The relationship of programming, problem solving and mathematics has been widely documented (see, for example, Hoyles, Sutherland and Evans, 1985; Noss, 1985) and will not be discussed here. There has, however, been an inclination to exclude programming from the computer activities of girls on the ground of girls' lack of motivation and poor performance in subjects such as computer studies which traditionally have included programming, usually in BASIC. We would argue that the exclusion of programming from the computer activities available to girls reinforces sex stereotyping of girls. Although some forms of computer interaction may leave the user feeling comfortable, if there is no feeling of *power* or *control* over the computer, which comes with some sort of programming, there will not be a sense of computer competence or confidence.

If we are to introduce programming, however, we must be careful to recognize that it is often seen as demanding a particular style of working. For example, Dirckinch-Holmfeld (1985) argues that 'data logics' are essentially alien to the female form of cognition because of social and cultural influences. Here 'data logics' means the translation of actions into a logical, linear and unambiguous language. She suggests that for such a translation to take place, the problem must essentially have already been solved. A mechanistic form of thinking which devalues the role of intuition during the process of problem solution may therefore be encouraged. If programming is viewed in this narrow way, that is, in a way which emphasizes planning and logical progression, it undoubtedly will alienate pupils with different preferred learning styles. Yet there *are* different, and equally valid, approaches to programming. Computer programming is not *necessarily* an activity which imposes a linear planning style, what is termed 'hard mastery' by Turkle (1984). Many programmers are 'soft' masters. Soft masters interact with the computer and work on a problem by arranging and rearranging the elements within it rather than carrying out a preworked plan. By this method they often come up with new and surprising results. It is the case that many expert programmers use this negotiating style whenever they tackle something new. The points to be made here are that there are different programming styles and girls may favour the more soft approach.

The programming language used obviously has a crucial influence on the style adopted by the programmer. The computer language Logo, because of its structured and procedural nature, is particularly encouraging to soft mastery. Logo can provide a new way of building a relationship with computational objects and formal systems since these objects can be named, explored and manipulated. However, even whilst using a computer language such as Logo, we must recognize the overriding influence of what is valued and assessed in the classroom. Carmichael et al. (1985), in reporting research on the use of Logo, found large variations in performance between boys and girls — in favour of boys. These performance variations were based on a series of test results and from a content analysis of the amount and variety of work saved. Although it was suggested that these results were primarily a function of the amount of time students were able to engage with Logo, the whole issue of the type of work which made up the performance criteria needs to be addressed. For example, in discussing variety of work, Carmichael et al. reported that the girls tended to stay with the ideas they knew and with which they were familiar. Girls were viewed in a rather negative way as being thorough, methodical, and aiming for perfection. In contrast, the work of the boys was reported in positive tones as being innovative and experimental. Boys, it was said, explored many more Logo primitives, were eager to make their own games (frequently combining text and graphics in the process) and discovered new ways of creating movement on the screen. Competence in programming was therefore assessed by acquaintance with a large number of commands, by the complexity of programs developed and by the completion of challenges. Depth of understanding and creativity in the multiple uses of a small number of commands did not appear to be valued so highly.

As acknowledged by Carmichael et al. in their concluding remarks, this type of assessment of programming skills can bias the appeal of computers. Programming which stresses rules and winning would tend to be incompatible with such socialized female values as relational ethics, as pointed out by Gilligan (1982). Thus if we emphasize challenges and individual mastery we must first recognize that girls may suffer negative emotional reactions (for example fear of success at the expense of others). More fundamentally, we must also acknowledge that we are *imposing* a cognitive style which emphasizes the separation of self from object. Papert (1986) argued that some individuals prefer to work by *identifying with* objects and contextualizing knowledge through this identification. These people prefer to break down the barriers between self and object in order to work inside the situation rather than 'plan, freeze the plan and then execute the plan'. We would argue that these different styles of working need not affect the quality of the result. Care must be taken, therefore, not to assess performance in ways that *assume a particular approach.* In addition we must look to the distribution of topics we choose to assess. In the particular area of computing it has been reported by Anderson that 'High school females performed better than males in *specific areas of programming* . . . such as problem analysis and algorithmic application . . .' and concluded that 'computer tests . . . should be carefully examined . . . for bias against women and minorities and for bias against non-mathematical computer work' (Anderson, 1987, p.50 our italic). We come now to the fundamental issue. We want to question the widely held assumption that the way to be a successful programmer and problem solver is to exhibit what are termed autonomous learning behaviours (Fennema and Peterson, 1985). Such behaviours emphasize independent mastery over formal knowledge. We shall argue that girls and women may find this style of working alienating. Yet they *are* quite able to think and talk about formal systems if a different approach is taken. We now turn to the results of our own research as illustration of these general points.

The Logo Maths Project

Overview

The Logo Maths Project (1983-6) (Hoyles and Sutherland, 1989) was a longitudinal study of the use of Logo within the mathematics curriculum in which systematic data was collected for four pairs of case-study pupils (aged 11-14 years). Two computers were placed in the mathematics classroom and pairs of pupils took turns to work with Logo during their 'normal' mathematics lessons. Pairs were chosen to take into account spread of mathematical attainment and the teacher's opinions as to constructive working partnerships. The research data included recordings of the pupils' Logo work, all the spoken language of the pupils whilst working with Logo (a video recorder was connected between the computer and the monitor), the researcher's interventions and a record of all the other mathematical work undertaken by the pupils. The video recordings were transcribed and these, together with researcher observations and teacher and pupil interviews, provided the basis for the research results. Since 1984 the research has been extended into ten further classrooms (the extended network) where the control of the Logo work was the responsibility of the teacher rather than the researchers.

In the Logo Maths Project attention was paid to social and contextual issues such as modes of working and cognitive style. No gender differences were observed in terms of motivation, setting of challenging goals, persistence, showing of initiative and anxiety about the computer as a machine. Girls enjoyed the collaborative atmosphere and excelled at 'pattern spotting' and seeing the structure underlying a program or problem. There was also some indication of a change in teacher perception of the girls. After the Logo work, girls were more likely to be seen as having potential for creative problem-solving. This may be because the girls actually changed their behaviour or it may be that the teacher merely saw them differently — in either way it is of significance.

In addition to the above general points, by the end of the three years of the project we found no evidence of any differences between girls and boys with respect to their:

- ability to use the ideas of structured programming when working on a well-defined task
- ability to use the ideas of variable
- facility with a top-down or bottom-up approach to planning.

We shall now discuss how the issue of gender *does* seem to affect the learning situation in terms of:

- different ways of interacting with the computer as a problem-solving tool
- the nature of collaboration and discussion between pupil pairs
- attitude to mode of working.

Ways of interacting with the computer as a problem-solving tool

(i) *Findings*

At the beginning of the Logo Maths Project, pupils were given the freedom to choose their own goals and develop their own problem-solving and programming strategies. Researcher interventions aimed to focus on process and to encourage the pupils to predict and reflect. We did not impose any 'idealized' problem-solving strategies on the pupils as we wanted to investigate the different ways pupils would work naturally at the computer. We carried out an analysis of the types of goals on which pupils chose to work. These goals were classified along a dimension concerned with the extent to which the pupils defined and planned the final outcome of their work. We labelled the extremes of this dimension, loosely-defined/well-defined goals. On the one hand, loosely-defined goals were characterized by a lack of detailed pre-planned structure: they evolved out of exploratory 'making sense of' activity. When pupils work on loosely-defined goals they build up their goal whilst interacting with the computer. The procedural nature of Logo encourages this creative building-up activity. An example of a loosely-defined goal is given in Figure 1. The final superprocedure (Starbuster) was created from two separate modules (SS and FS). As the goal emerged the two pupils used four levels of nested subprocedures. They also for the first time defined procedures for the interfaces between the named graphical objects. Well-defined goals on the other hand have a well worked out and pre-planned overall structure and global product. When working on well-defined goals pupils know at the beginning of the session what they aim to produce by the end. George and Asim's butterfly project is an example of a well-defined goal (Figure 2). The picture had been carefully drawn out on graph paper before interaction with the computer and it was executed on the computer exactly as planned.

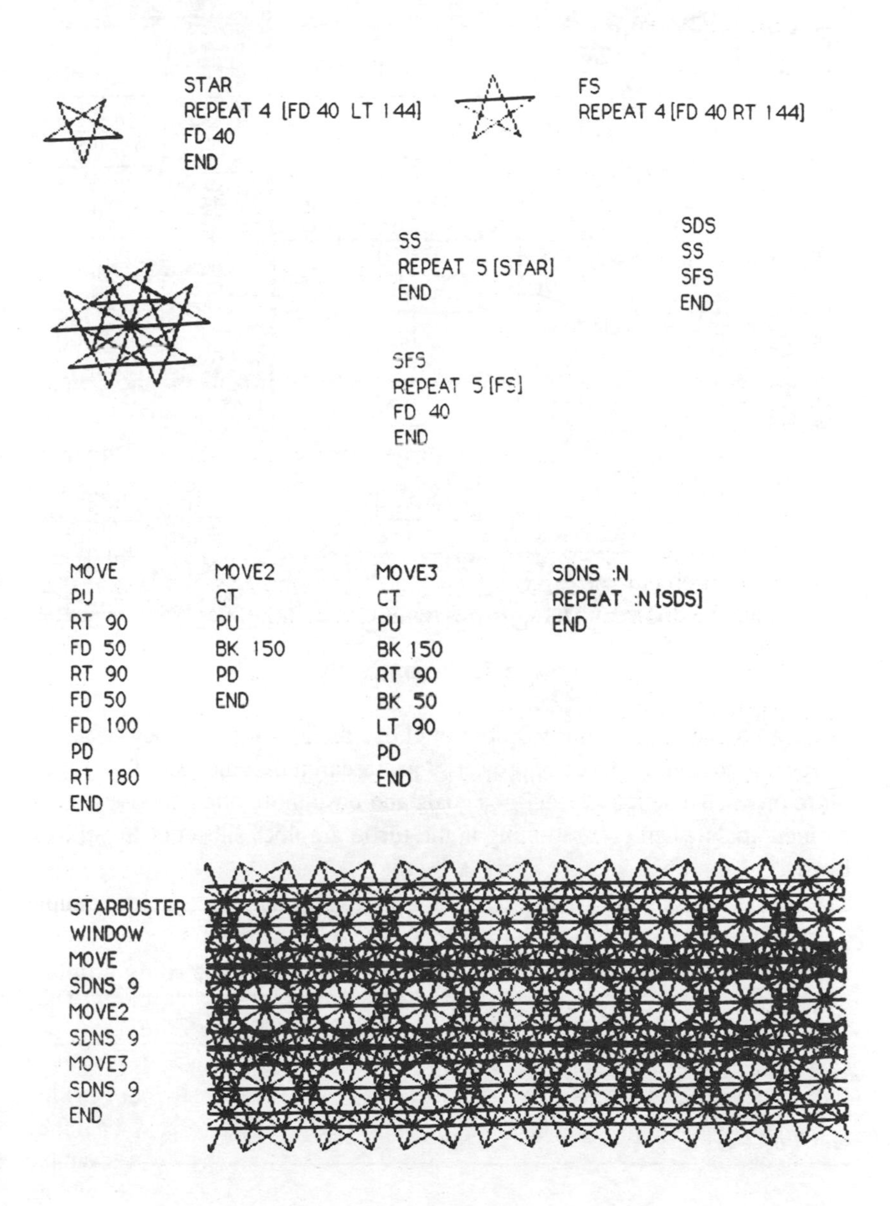
STAR
REPEAT 4 [FD 40 LT 144]
FD 40
END
FS
REPEAT 4 [FD 40 RT 144]
SS
REPEAT 5 [STAR]
END
SDS
SS
SFS
END
SFS
REPEAT 5 [FS]
FD 40
END
MOVE
PU
RT 90
FD 50
RT 90
FD 50
FD 100
PD
RT 180
END
MOVE2
CT
PU
BK 150
PD
END
MOVE3
CT
PU
BK 150
RT 90
BK 50
LT 90
PD
END
SDNS :N
REPEAT :N [SDS]
END
STARBUSTER
WINDOW
MOVE
SDNS 9
MOVE2
SDNS 9
MOVE3
SDNS 9
END

Figure 1: 'Starbuster'

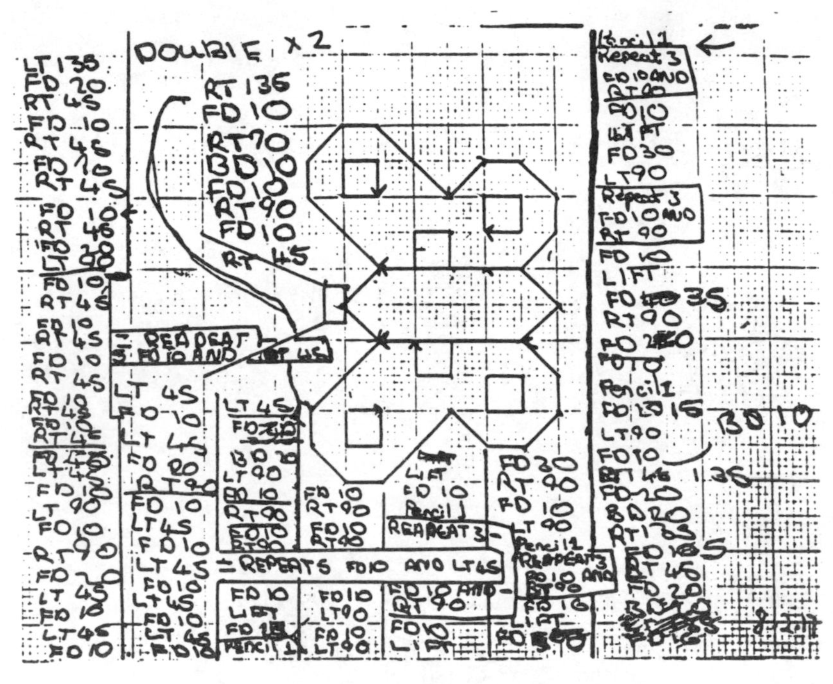

Figure 2: the butterfly

When pupils are given the freedom to chose their own goals they exhibit preference for certain types of projects. Our research data indicates that girls more often choose loosely-defined goals and boys more often choose well-defined goals when programming in the turtle graphics subset of Logo, as illustrated in Table 1.

Table 1:
Ratio of loosely-defined to well-defined goals for the case study pupils

	Loosely Defined	:	*Well Defined*
Girl Pairs	22	:	1
Boy Pairs	5	:	21
Girl/Boy Pairs	14	:	11

We suggest that in our school culture the more well-defined a goal appears to be, the more value it is accorded. School culture reflects our production-oriented society. Girls' apparent lack of emphasis on product can easily lead to their work being undervalued. This was mentioned earlier with reference to the Carmichael et al. (1985) project where it was reported that 'without exception the boys saved more programs than the girls'. We would argue that this focus on product and quantity not only reflects a bias in performance criteria but also would tend to emphasize a certain type of learning. Carmichael described how the girls tended to stay with the ideas they knew and with which they were familiar, giving as an example one pair of girls who spent most of the first year working with circles, 'filling them with different colours and combining them to make lollipops and balloons'. We, too, have observed a similar phenomenon but do not make the same negative value judgement of such explorations since they are important for certain aspects of learning. For example, we have evidence that the girls, through their extensive work on loosely-defined goals, were more able later to break down well-defined tasks into modules. Well-defined projects, on the other hand, were more likely to provoke pupils to attend to the detail of turtle turn and make links between angle and turtle turn (Hoyles and Sutherland, 1986a). This point is again echoed by Carmichael et al. who found that boys were more likely than girls to know the size of the angle needed in a turtle graphics project. We would interpret this result as a consequence of the type of activity experienced.

All the extended network pupils were given paper and pencil Logo tasks at the end of the project. From one of these tasks (the K task, Figure 3) we again have evidence that the boys were more likely to have made the links between turtle turn and angle. Almost twice as many boys as girls gave the correct RT 135 answer for the 'K' task. In the 'M' task (Figure 4) more girls than boys modified the correct command, that is, understood the sequence although making an incorrect modification. There is also evidence from these tasks that the boys are focusing more on the object itself and are not so much concerned with the initial orientation of the turtle before it starts to draw. In the 'H' task, which involved asking the pupils to draw a figure from a list of commands, 11 per cent of boys as opposed to 3 per cent of girls drew the correct figure without bothering with initial orientation. Similarly in the 'F' task, which involved asking the pupils to write a procedure to draw the letter F, 14 per cent of boys as opposed to 3 per cent of girls give a correct solution apart from the initial orientation of the turtle. In contrast girls seem more likely to focus on the relationship between objects on the screen and are more likely to define subprocedures for the interface commands which join two graphical objects. More research is needed in order to substantiate these claims.

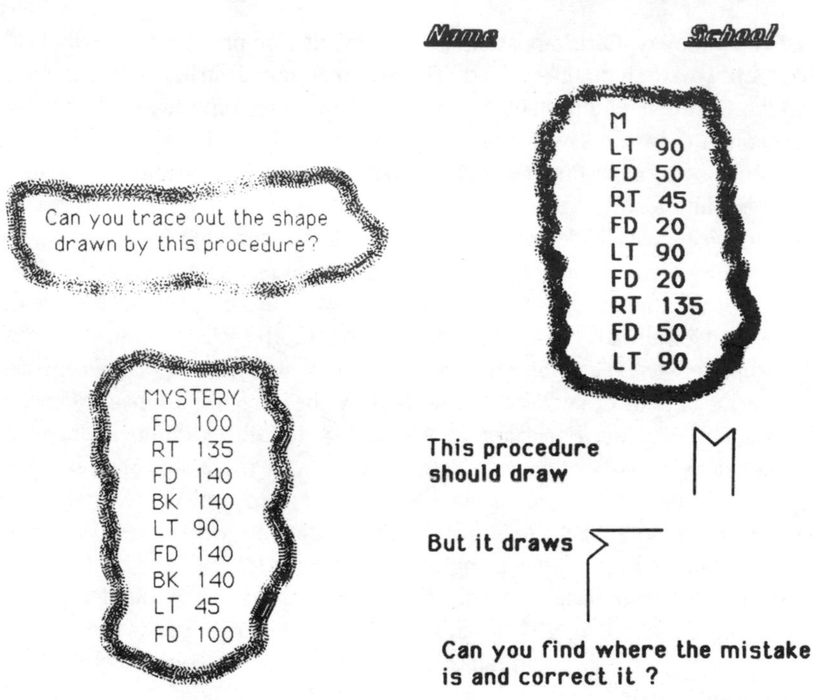

Name
School
Can you trace out the shape drawn by this procedure?
MYSTERY
FD 100
RT 135
FD 140
BK 140
LT 90
FD 140
BK 140
LT 45
FD 100
M
LT 90
FD 50
RT 45
FD 20
LT 90
FD 20
RT 135
FD 50
LT 90
This procedure should draw
But it draws
Can you find where the mistake is and correct it ?

Figure 3: the 'K' task

Figure 4: the 'M' task

(ii) *Some examples*

By the end of the first year of the Logo Maths Project we became aware of these differences between the pupils in their choice of goals. We therefore gave tasks to encourage pupils to choose more flexible types of work, as the following examples illustrate.

George and Asim are two of our case study pupils. Throughout their first year of learning Logo they always chose for themselves well defined picture goals. They pre-planned their work very carefully, usually away from the computer. Their planning took the form of drawing out their design on graph paper, writing a linear sequence of commands and splitting these into sub-procedures only when this was imposed by the storage restrictions of the machine (see, for example, Figure 2). They never worked in an experimental way with sub-procedures and did not come to appreciate the intrinsic nature of turtle geometry; that is that the same 'shape' in a different position or orientation can be defined by the same procedure. This absence of 'hands on' exploratory activity was detrimental to their understanding of the ideas of structured programming. The pair never built up structured programs so they found it difficult to break down a well-defined procedure in a structured manner. At the beginning of the second year of the project, we decided to encourage George and Asim to try a less well-defined task at the computer. We gave them a set of 'abstract' images allowing them to choose which image to draw and how they would go about it. We hoped that working on one of these loosely-defined designs would encourage a different way of interacting with the computer. They decided to draw a spiral image (Figure 5). In contrast to their previous work, it was not regarded as a 'picture of anything'. Although apparently well-defined at this stage, it turned out to be just a starting point and the session developed in a delightfully unexpected way.

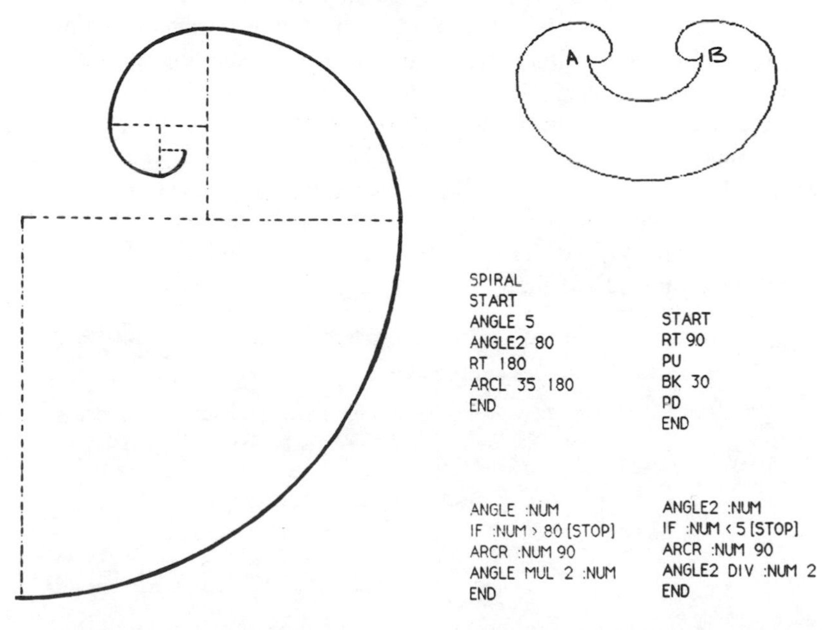

Figure 5: spiral image (George and Asim)

Figure 6: from spiral image to semi-circle (George and Asim)

The pair built up their spiral by a sequence of quarter circles with radii increasing in a geometric progression (multiplying by 2). When putting this into a procedure they were introduced to recursion. They then decided to 'reverse all the procedures' so that 'they could have another one coming round here'. They managed to adapt their recursive procedure to achieve this. Finally they decided to join up A and B with a semi-circle (Figure 6) and had to spend considerable time and experimentation working out how to do this! It was this final step that led them to look at the global structure of their project and indeed recognize more explicity that the first input to the ARCR command was the radius and not the diameter! During this session the pair had felt free for the first time to explore mathematical and programming ideas and build on their initial module in an exploratory manner.

Sally and Janet, two other case study pupils, preferred loosely-defined goals and during their first year of programming in Logo ten out of twelve of their own projects were loosely defined. Although Sally and Janet chose themselves loosely-defined goals, when they were given a well-defined goal to work towards they showed that they could make a plan and use the ideas of structured programming. We asked them to draw a pattern of squares (Figure 7). The pair initially analysed the task as consisting of four outer squares (before working at the computer). They therefore defined a square module, and, in trying this out decided to delete the last command (see Figure 7), so that the four outer squares would be produced by typing SQUARE SQUARE SQUARE SQUARE. They then added the commands for the inner square to their superprocedure PRISON. The act of making a written record of these commands led them to notice that they could also by adding a interface command use their SQUARE procedure for the inner square.

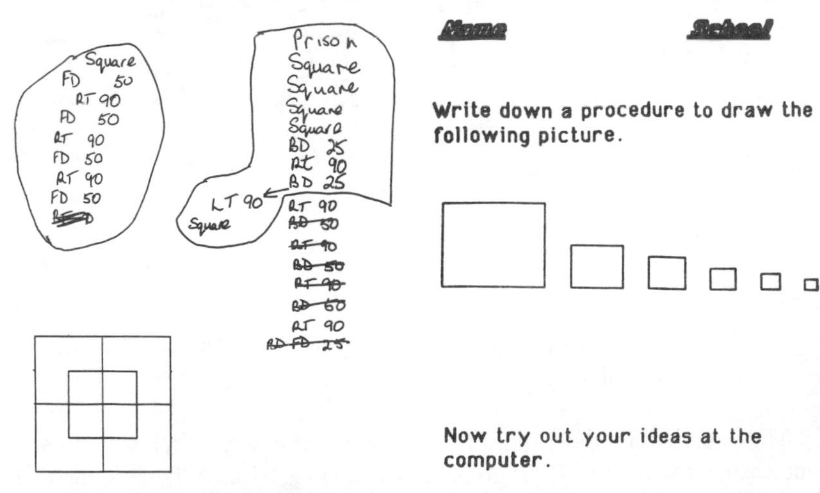
Square
FD 50
RT 90
FD 50
RT 90
FD 50
RT 90
FD 50
Prison
Square
Square
Square
Square
BD 25
Rt 90
BD 25
LT 90
Square
RT 90
RT 90
Write down a procedure to draw the following picture.
Now try out your ideas at the computer.

Figure 7: pattern of squares
(Sally and Janet)

Figure 8: the decreasing row of squares task

(iii) *Difference in approach to the same well defined problem*
Throughout the Logo Maths Project we occasionally gave the case study pupils, either individually or in pairs, the same well-defined task and we observed differences in programming style between the girls and the boys. These differences cannot be adequately described by reference to the dimension of top-down planner and bottom-up planner but are more to do with mode of computer interaction. In fact one boy and one girl, Asim and Sally, both tended to be top-down planners whereas George and Janet both tended to be bottom-up planners. In contrast to Asim, though, Sally always wanted to work initially in direct mode. Her behaviour masked the fact that she nearly always started a project with a clear top-down plan. Our evidence for this came either from her written record or from her interactions with her partner, Janet.

All the case study pupils were given the decreasing row of squares task (Figure 8).

Sally and Asim both made top-down plans but, whereas Sally tested all the modules of her top-down plan and then used these to build up the row of decreasing squares *before* defining the final superprocedure, Asim defined a superprocedure *straight away in the editor.* He then had considerable debugging problems because he had not attended to state and interface details in his square module. Similarly, when Sally and Janet were working together on a well defined task, they consistently worked in a way which involved testing individual modules and building these into the final product *before* defining the superprocedure. The fact that they did not start the project by defining the superprocedure did not mean that they did not have a top-down plan of the problem solution. When given the same task Sally and Janet, unlike George and Asim, used 'hands on' activity as a way of getting into the problem. Once involved in the problem, they took time off to discuss their global plan, whereas George and Asim discussed their global plan before typing any commands into the computer. There is the danger that superficial observation could lead to the conclusion that Sally and Janet were not planning. Our evidence suggests that they did plan when working towards well defined goals but the nature of their interaction with the computer was different from the boys. They used interaction with the computer to get started and to engage in the problem. Table 2 illustrates the comparisons described above by reference to the type of programming activity undertaken, the number of utterances of the pupils and the number of computer commands used.

Table 2: Task analysis for arrow
(a) Sally and Janet

Type of programming activity	*No. of utterances*	:	*No. of computer commands*
Planning	2	:	0
'Hands on'	27	:	12
Planning	44	:	2
Defining a procedure	10	:	24
'Hands on' (Trying out of procedure which works)	3	:	6

(b) George and Asim

Type of programming activity	*No. of utterances*	:	*No. of computer commands*
Planning	22	:	2
'Hands on'	10	:	8
Refining plan and negotiating syntax	51	:	0
'Hands on'	12	:	1
Defining a procedure	7	:	6
'Hands on' (Trying out of procedure which has a small bug)	4	:	1
Debugging	1	:	1
'Hands on' (Trying out of procedure which works)	8	:	9

These differences observed for our case study pupils have also been found in the extended network data. Boys were more likely than girls to 'jump in' at too high a level, in terms of defining a procedure in the editor when they did not have a clear idea of the process within the procedure. This often led to difficulties in debugging. For example all the extended network pupils were given the task of drawing a row of identical four squares at the end of their first year and second year of Logo programming. Analysis of the data showed that whereas there was no difference between the number of boys and the number of girls who obtained a working program, there were considerably more boys than girls who wrote procedures which they then could not subsequently debug.

(iv) *Pupils' choice of variable names*

Another unexpected difference between the case study pupils appeared when the pupils' choice of variable names was being analysed as part of a study to investigate the pupils' developing use and understanding of variable in Logo (Sutherland, 1988). It had been observed at an early stage in the research that pupils were attaching too much significance to meaningful variable names like SIDE and NUM. Using a wide range of variable names including 'nonsense' names was considered to be an indication that pupils understood that in Logo *any* variable name could be used. In analysing the range of variable names used by the pupils it was found that not only did girls choose to use a much wider range of names but they were also more likely to be imaginative in their choice of names. The names chosen by George and Janet throughout the three years of the project are given below to serve as an example of this point.

Janet: NUMBER, NUM, SIDE, SCALE, N, D, E, K, M, J, X, R, YT, HT, RAD, TWA, ONE, TWO, JACK, JILL, CAT, WO, MAN, NUT, IT, ORE, JOE, LEE, PIG

George: NUM, SCALE, LEN, LEN2, NUT, SIDE1, SIDE2, AREA1, AREA2, DISTANCE, ANGLE, X, Y, B, O, DIG

When Janet solved the 'Arrow' task (Figure 9) she used the variable names JACK and JILL and when George solved the task he used the variable names NUT and NUM. Both solutions were equally sophisticated from both a programming and a mathematical point of view.

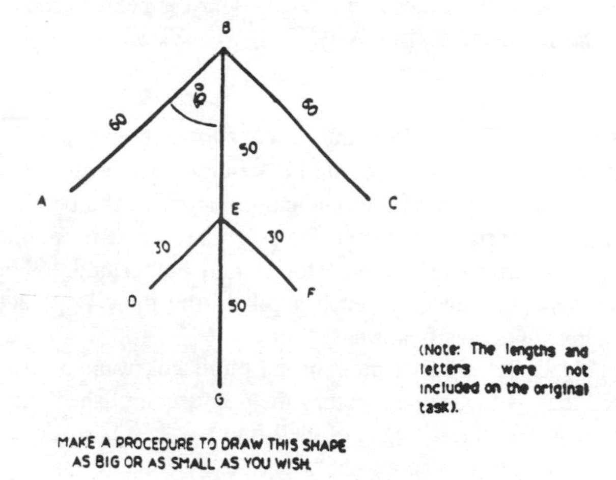

Figure 9: the arrow task

There was also evidence that being able to use any variable name (and any procedure name) provided a motivating factor for the girls when they were working on a task which involved defining simple functions in Logo. The following are examples of functions defined by girl pairs:

```
TO HAZEL :NUT                 equivalent to x->x/3
  OUPUT :NUT/3
  END
TO POXIE :SEAN                equivalent to y->y-17
  OUTPUT :SEAN-17
  END
```

The imaginative choice of variable names in no way changes the mathematical structure of the procedures and quite obviously contributed to the pupils' engagement in the task. The boys did not appear to want to use variable names in the same way and were more likely to write procedures in the following form:

```
TO ASIM :P                    equivalent to z->3z
  OUTPUT 3* :P
  END
```

We are not attempting to put any differential value on either nonsense names (for example SEAN) or abstract names (for example P) but it is worth considering that in the 'normal' mathematics context pupils are usually restricted to abstract names. This study shows that for the girls the more free choice of variable names in the Logo context appears to add a motivating factor to the mathematical activity.

The nature of collaboration and discussion between pupil pairs

A classification system for the pupil discourse was developed in order to obtain an overall picture of the qualitative nature of the peer interactions, to facilitate comparisons between the pupil pairs and to monitor changes in interaction patterns over time (Hoyles and Sutherland, 1986b). All the verbal 'on task' utterances of each pupil during their Logo activity were coded, using this classification system.

This analysis showed that most of the pupil talk was specifically action orientated, that, is focused on 'getting the task accomplished'. Little attempt, particularly at the initial stages of the Logo work, was made by the pupils to explain or convince one another of what was meant or why a proposed course of action should be taken. This was observed for all the case study pairs, but particularly was marked for boys. As the project progressed there was a marked increase in more extended interactions and attempts to communicate strategies and explain predictions. However, gender differences were observable with a girl more likely consciously to share with her partner her representation of the problem and her ideas for problem solution. The differences found in our overall analysis are very similar to those of Nielsen and Roepstorf who reported that:

> The girls tend to orient themselves towards each other. The human relation seems equally — or more — important than the object they are working with. They seem to operate within a we-circle, and the co-operation is predominantly collective. The boys apparently work more individually, and orient themselves primarily towards the object, and secondarily, but also essential towards each other (sic). Though with attitudes which are different from those of the girls. The boys also co-operate, but the dominant tendency is towards an individual performance, often with what we have termed an element of competition . . . Whether or not the individual girl was capable of solving the posed problem, or knew the next step, she most often drew the other girl into a collective co-operation by using verbal suggestions like: 'Don't you think . . .?' or 'What if we turned it more?' Whether or not the other girl agreed, the suggestion was very often made the object of a discussion, in which mathematical concepts were used only tentatively. (Nielsen and Roepstorf, 1985, p.65)

Thus girls were less likely to fight for control than boys who often seemed concerned to establish their autonomy and impose their problem representation and solution. Boys used few verbal supports of their partner's contributions. They appeared to be trying hard to convince each other which led to a competitive style of speech. Boys tended to be more careful with the exact local detail of their designs, which were often very carefully planned in advance. Their interactions were therefore often simply suggestions for actions which were not negotiable as they had already been worked out in advance. This type of verbal interaction is obviously related to the boys' tendency to choose well-defined goals. It might be easy to interpret from the boy's language that they do not collaborate or listen to each other's suggestions since their final actions often appeared to be based on who was the strongest or most persistent. In fact we have found frequent cases when ideas from a partner were taken up later, but not acknowledged!

Let us take some selected interactions during work on the Arrow Task, (Figure 9) described earlier, as an illustration of the above points. We shall focus on the processes by which the pupils, working collaboratively, negotiated how they saw the task and developed a plan as to how it might be achieved. For both pairs the coding of the pupil utterances indicated a high overall level of elaborated argument.

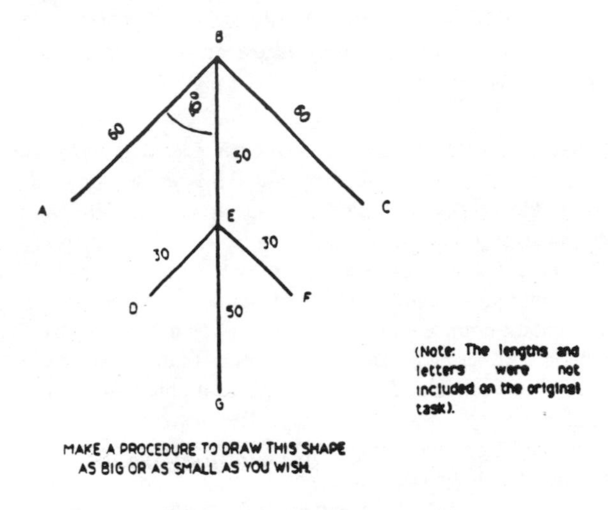

Figure 9: the arrow task

Sally and Janet, the two girls, produced the figure in direct mode (their preferred working style). The figure they drew did not reflect the ratios given and AB was made to be equal to BE and twice as big as DE. At this point there was a detailed discussion of how the procedure should be built, including elaborated arguments in the attempt to come to a consensus. The girls spontaneously saw that they could use the same input for AB and BE but wanted at the outset to use another input for DE. Janet explained her plan using language which was context embedded, that is by reference to a particular example:

> Alright . . . we work it out 'cos that will have to be something called JACK and that JILL if you get what I mean . . . all the 50s then the 25s

They started to build a procedure:

```
HILL :JACK :JILL
RT 90
BK :JACK
```

At this point Sally intervened:

S Wait a minute you have to do . . . no BK MUL . . .

She wanted to operate on an input and tried to elaborate why.

S . . . Em you say for this one you say BK :JACK and for this one you multiply it by two 'cos that's half

Janet saw the structure but believed *two* numbers had to be used as inputs. Her argument was again very context specific.

J No 'cos listen, look. But anyway say that's 100 and we put in 100 then that would do that a 100 but you'd have to put in another number that . . . So instead of putting in two numbers. Listen.

Sally seemed to become more confident in her idea in the face of Janet's conflicting perspective. She tried to justify her proposition and in so doing provided Janet with some 'scaffolding'. The pair discussed, they listened to each other and responded to the context arguments. They decided eventually on a plan, suggested by Janet, that they could get rid of one input JACK.

S But we're not going to put any old number in 'cos it won't be the same pattern. Look . . .

J Yeah, but if you put in 75 then they're not going to be 75, they're going to be any old number.

S Yes, that's why we're going to multiply it.

J Yes, but you don't need to multiply it, that's what I'm saying if you say, em, if that one say that times by two it would be that wouldn't it?

S Divided by two.

J Yes, you know what I mean.

S Alright.

J But I don't know how we're going to do it. We can get rid of JACK.

It was apparent that George and Asim, when given the same task, saw it differently. George saw it as being made up of two parts, ABCE *placed on top of* DEFG, while Asim saw it purely as one arrow superimposed on another half its size. From the first section of the transcript, it is clear that neither boy was willing to change his representation and accommodate his partner's view. So they essentially work out individual plans, which then come into conflict when the pair had to agree on a Logo program. In addition, the figure the boys worked towards did not reflect the ratios of the lengths given but were the same as those perceived by the girls.

The boys drew DEFG in direct mode. With the turtle at E the discussion showed that they had not agreed on an overall view of the task. The nature of their discussion did not help them to achieve a consensus.

A Just draw an arrow and MUL it.

G Go forward the same distance.

A But you only need one.

G No, these are the same distance.

A What? That's the small one.

A I know but you only need one arrow, then you can MUL it as you like.

George apparently agreed and said 'Oh, o.k.'. After some off-task talk, however, he again suggested a move forward — his original plan.

G FD 20 'cos that's always the same.

A No it isn't.

G It is.

A No it isn't.

G It is 'cos the that's the same distance as that.

A Yes, but if we . . .

Even though there was no consensus George started to type FORWARD . . . Asim reverted to an authority figure, i.e. the teacher, to try to settle the dispute.

A Go on ask her if what we are going to be drawing afterwards is that or the whole thing.

G I know 'cos that is always going to be the same.

A Which one?

G That FD 20.

A I know, that's what I mean. That's why I have to check up.

G It is always the same. I know that.

G The FD 20 'cos that is always the same distance.

A I know but it depends on whether you're going to be drawing, if you're going like that or if you're going to be drawing an arrow there, an arrow there . . .

G It's still always the same.

A If you're wrong I'll kill you. Go on.

This style of interaction was reflected throughout the session. Disagreements tended to be resolved by reference to an outside authority. The boys, unlike the girls, also spent a lot of time discussing syntax but again could not agree and eventually asked the teacher.

A BK two dots MUL.

G No it's not. No BK NUM.

A 10 no you don't put MUL you only put NUM it's got nothing to do with MUL. It'll be NUM two dots 10.

G No it's backward 10. Multiplied by 2 each time. No it's BK two dots NUM.

A That's what I keep saying, why don't you listen to me and then you put 10 afterwards or 10 before it.

G Ask Miss.

Even though George continued with his idea, Asim was still not convinced and when they came eventually to write a procedure definition, he again remarked:

A What I want to understand OK is that is this arrow supposed to be one arrow like that . . . forget about the top or is it supposed to be like that or is it supposed to be like that?

G Like that.

A Or is it supposed to be a whole thing?

G Yeah. That that that that . . .

A They're supposed to be together.

As mentioned earlier the boys *did* take on each other's ideas, although their language and combative style made it seem that they did not. This is illustrated in the following extract. George pointed out at the beginning of the episode that the initial orientation of the turtle was not important (as the Arrow was to be a completely general procedure).

G Actually we can't have that LT 90 there because if we do that every time we do arrow it will go . . .

Much later when the boys were at the point of building their procedure, it was Asim who took up this point. They had entered the editor when Asim said:—

A What is it LT 90? There isn't any point in turning round is there.

G What I'm just turning it upwards.

A There's no need to do that remember.

It turned out, however, that George was planning ahead — although he had not yet disclosed this to Asim! He realized that, given the way they had constructed their program (in a modular way with two calls of ARROW with a centre CT in between; see below), they had to think of turtle orientation after all!

G There *is* a need . . . if you do that and centre it again you have to turn it again.

They therefore introduced an input, NUT, to their procedure which oriented each call of ARROW in the same way:

```
TO ARROWPLUS :NUT
LT :NUT
ARROW 20
CT
LT :NUT
ARROW 40
END
```

Attitude to mode of working

An attitude to mathematics questionnaire was administered to all pupils (total 222) in the extended network schools in order to explore the effects of working with Logo. The attitude questionnaire was administered twice in fairly quick succession (to get some idea of stability of response) and thereafter on a termly basis. The responses to the questionnaire were coded and computerized and an analysis undertaken using the SPSS package.

For the purpose of this chaper we shall focus on two of the results from this questionnaire data. In response to the item 'I enjoy working with others in the class' the responses of the girls over the period of the research became increasingly more positive. Initially the response of the girls to this item was less positive than that of the boys while at the end of the project the reverse was the case.

The response to the following item also showed interesting gender differences:

'When I get stuck on a maths problem:

A. I like to try and work it out for myself.

B. I ask someone else in the class to help me.

C. I ask the teacher.

D. All three on different occasions'.

Boys consistently were more likely to state a preference A, for 'working it out on their own'. Very few boys ticked preference B. In contrast girls were more ambivalent in their responses, but exhibited an increase in the response B over the period of the research.

In interviews with the individual pupils, preferred working mode — individual or with other — was probed further. Girls who expressed a liking for group work emphasized the importance of receiving help from their peers. Boys in contrast saw group work as distracting from their individual achievements. Boys who preferred individual work focused on the individual challenge involved and the opportunity for being able to puzzle things out for themselves without arguments from peers. Girls who liked individual work in contrast emphasized how they could progress at their own pace.

Conclusion

In conclusion, we wish to reiterate our recognition that, by our focus on gender specific behaviour and attitudes, we are ignoring important individual variations. Nonetheless, by adopting this focus, we have been able to distinguish different styles of learning and modes of interaction with the computer. In our analysis, we have been aware of the danger of coming to quick and superficial conclusions. Consideration of individual episodes, or short-term performance results, can sometimes be misleading, and it is all too easy to come up with slogans such as 'Girls do not plan', 'Boys do not collaborate in their computer work', and 'Boys are better than girls in programming'. We have confidence in our interpretations because they are backed up with systematic, longitudinal data and we have found, not surprisingly, that the situation is far more complex than this: girls *do* plan and boys *do* collaborate, but not necessarily in the way we would predict. Who achieves more is at least partly dependent on how we organize and assess the activity.

References

Anderson, R. E. (1987), 'Females surpass males in computer problem solving: findings from the Minnesota Computer Literacy Assessment', *Journal of Educational Computing Research,* Vol.3., No.1.

Carmichael, H. W., Burnett, J. D., Higginson, W. C., Moore, B. G. and Pollard, P. J. (1985), *Computers, Children and Classrooms: a multisite evaluation of the creative use of microcomputers by elementary school children.* Ontario, Canada: Queen's Printer.

Dirckinch-Holmfeld, L. (1985), 'Information technology and cognitive processes', *Proceedings of Third International GAST Conference.* London, April.

Fennema, E. and Peterson, P. (1985), 'Explaining sex-related differences in mathematics: theoretical models', *Educational Studies in Mathematics,* Vol.14, No.3.

Gilligan, C. (1982), *In a Different Voice.* Cambridge, Mass: Harvard University Press.

Hoyles, C., Sutherland, R. and Evans, J. (1985), *A Preliminary Investigation of the Pupil-Centred Approach to the Learning of Logo in the Secondary School Mathematics Classroom.* London: Leverhulme Trust.

Hoyles, C. and Sutherland, R. (1986a), 'When 45 equals 60,' *Proceedings of the Second Logo and Mathematics Education Conference.* London: Institute of Education, University of London, July.

________ (1986b), 'Peer interaction in a programming environment', *Proceedings of the Tenth Psychology of Mathematics Education Conference.* London: July.

________ (1989), *Logo Mathematics in the Classroom.* London: Routledge (in press).

Nielsen, J. (1987), '"This is a very predictable machine", said Michael — on computers on human cognition', in R. Gregory, (ed.), *Creative Intelligence.* London: Pinter.

Nielsen, J. and Roepstorf, L. (1985), 'Girls and computers — delight a necessity', *Proceedings of Third International GAST Conference.* London, April.

Noss, R. (1985), *Creating a Mathematical Environment through Programming: a study of young children learning Logo.* London: Institute of Education, University of London.

Papert, S. (1986), 'Beyond the cognitive. The other face of mathematics: lecture notes', *Proceedings of the Tenth International Conference of the Psychology of Mathematics Education, Plenary Lectures.* London.

Sutherland, R. (1988), 'A longitudinal study of the development of pupils' algebraic thinking in a Logo environment', unpublished PhD thesis. London: Institute of Education, University of London.

Turkle, S. (1984), *The Second Self: computers and the human spirit.* London: Granada.

Sixth-Form Girls Using Computers to Explore Newtonian Mechanics

Susan Burns and Teresa Smart

Girls were not the subject of this investigation. In our planning we wanted to look at how the teaching of certain aspects of the A-level curriculum could be facilitated by using a computer and, in particular, whether Logo could be helpful in developing understanding of mechanics. However, when we introduced the work with mixed classes of A-level mathematics students, it was, in the majority of cases, the girls who showed interest and volunteered to continue the work in extra sessions. We were happy to work with the girls on their own. This gave us an opportunity to observe the particular ways in which the girls worked with and related to computers. Although it is stated that 'Generations are now coming into the sixth form many of whom can program and enjoy it' (Mathematics Association, 1985) we question what proportion of these are girls. From our own experience as teachers, we know that quite a number of students, and in particular girls, are entering A-level courses with no experience of working with computers and are also anxious about the prospect. At the 1987 sixth form girls conference, 'Maths and your future' (organized by Gender and Mathematics Association (GAMMA) in conjunction with Goldsmiths College) the most popular mathematics workshop was 'Logo for beginners'. All the workshops' participants were girls studying A-level mathematics who had no or very little computer experience.

Working with these girls, we wanted to use micros in A-level classes in ways that would: not assume prior knowledge of computer use; recognize that they might have limited confidence with computers; and encourage them to be more open to computers at the same time as enhancing the learning of mathematics.

We worked with eight 16-17 year old girls using the computer and the Logo programming language. These girls were all in the first year of an A-level mathematics course. More than half of them had done little or no work on computers. In this chapter we look at one aspect of this work: The Newton Microworld. This is a Logo-based modelling tool by means of which the movement of the turtle under Newton's Laws can be simulated and thus investigated. In particular, the motion of a particle in a frictionless world can be modelled. While using the Microworld all the students we worked with exhibited a conflict between their intuitive view of physics and the Newtonian laws that they had learnt in their mathematics or physics lessons. This we look at in more detail using case studies of two of the girls: Angela and Sadia (for fuller details see Burns, 1988).

We did not probe why it was that, on the whole, only girls responded to our request to join the extra computer-based sessions. We did, however, note that many of the girls seemed to have come because they felt some anxiety about their mathematics. This was even the case for those who were doing very well in their A-level classes. Many of the girls found their mathematics harder than their other subjects and hoped that these extra sessions would help. The others came because they enjoyed mathematics and wanted to take up the opportunity to explore one aspect further, especially given the opportunity to use computers in an environment perceived to be supportive. It is perhaps worthwhile to ponder why it was that the boys who were also finding mathematics harder than their other subjects did not feel interested in coming to these extra sessions. Possible answers from our experience would arise from the boys, even at the same level of achievement, feeling a lower level of anxiety. Many more boys were studying A-level physics and were already familiar with computers and maybe did not see the need to use a computer to explore Newtonian mechanics.

The Newton Microworld

The Newton Microworld is an extension of the Logo language written for Nimbus computers. The commands in the microworld allow students to experiment with and to model aspects of the real world such as: bouncing balls; objects colliding; projectiles; the effects of friction and circular motion. There is no need for any prior computer experience before working on the microworld. For example, a simple experiment might take the form:

```
setdir    30      — the turtle starts moving across the screen with
setspeed  10        speed 10 on bearing 030.
setmass    1

setforce  10 180  — the turtle is subjected to a single, one-time unit
kick                application of the force equivalent to an impulse.

pr speed          — enable the student to investigate the effect of the
pr dir              kick
```

We found that students who had not previously worked with Logo did not face difficulty learning to use the commands.

Background to the girls who worked with the microworld

The eight girls were self-selected and they came from a wide spectrum of A-level students. Some were studying physics at A-level and several were taking mathematics along with arts subjects, such as music and history. Two, Christa and Emma, were intending to go on to study mathematics at university. They all talked about enjoying mathematics, although most found it a difficult subject. Sadia said she found 'maths harder than my other two subjects (chemistry and biology) and rather frightening'. Nazneen commented that she was enjoying her mathematics course, but agreed with the word 'frightening'. Emma felt her mathematics was 'fair', but went on to say, 'Sometimes I feel unsure depending on the topic, particularly in applied mathematics'. Only two of the girls had a computer at home. This contrasted with their male peers. In Angela's class three of the boys' families each possessed more than one computer. Many of the girls were completely 'untouched' by the existence of computers. Sadia had never worked with a computer or programmable calculator before. Diana and Rehana had their A-level mathematics classes in a room furnished with three Nimbus computers but these were completely monopolized by the boys in the class. The computers were often used by these boys before the lesson, or during the break. It became clear, on interview, that Diana and Rehana had decided that these computers were not for them. It was almost as if the machines did not exist, even when they were faced with the colourful pictures, noise and dancing screen every time they entered their classroom.

All of these girls, whatever their previous fears and attitudes to computers, worked with great concentration and obvious enjoyment when engaged in the Logo Newton Microworld. They were able to build and change their own procedures, make models and then test them out. They coped admirably (sometimes better than us) with the inevitable frustration of working on a computer; for example, with messages such as 'fatal error', and with problems with the Nimbus operating system which would cause the keyboard to go dead when the network was overloaded (this problem has been overcome in the new version). The frustrations they did show were not with the computer but rather with their mathematical problems and misconceptions.

Sherry Turkle's comment that children give the computer 'the pyschological attributes of intelligence, feelings and morality' (Turkle, 1984) was borne out by our experience. For example, Angela stated that 'the computer has gone crazy' when the turtle responded to her orbit procedure in a way totally unlike what she was expecting. Amrit several times accused the computer of cheating. For example, on one occasion the turtle was moving with a speed of 20. It was given a kick at right angles to its direction of motion to increase its speed in that direction by 2. Amrit predicted that its resulting speed would be 22 and discovered it to be 20.1:

Amrit: When you kicked it the force wasn't totally applied.

Interviewer: You think the computer is cheating again?

Amrit: Yeah, I think it is. I mean, if you use f–ma the velocity should be 22 by now. The force was 10, the mass was 5 so the accelerations was 2. It hasn't accelerated by 2, only by 0.1.

This identification with the personality of the machine seemed important for the girls. When they felt confused it was more natural for them to ascribe characteristics of 'cheating' or 'craziness' to the computer than analyze what had gone wrong. With encourgement, the analysis came later.

Some misconceptions and how they were dealt with

Before working on the microworld, all the students were given an initial interview to probe their understanding of the mechanics of a particle moving under gravity. After working on the Microworld, for 12 to 16 hours, they were interviewed again. At all times during the computer work the girls were encouraged to explain how they were feeling about the computer and their understanding of the mechanics.

From their interviews it was clear that all the girls had many misconceptions about the motion of a particle under gravity or in an orbit. These misconceptions were strongly held and intuitively believed, although at the same time they could quote Newton's Laws learnt in the mathematics lessons. They believed that if you apply a kick to an object it will move off in the direction of the kick, even if the object already had a velocity. This is in contradiction to Newton's Laws. Andrea diSessa (1982) has shown that this is quite a common phenomenon. He observed a group of elementary school students learning to control a computer-implemented Newtonian object, a dynaturtle. 'These students revealed a surprisingly uniform and detailed collection of strategies, at the core of which is a robust "Aristotelian" expectation that things should move in the direction they are last pushed', even when the object is moving, not stationary, when the kick is applied.

The expectation is still held even by those with an understanding of classroom physics. As diSessa comments, on observing the interaction of an MIT physics undergraduate with dynaturtle,

> she could not relate the task to all the classroom physics she had. It is not that she could not make the classroom analyses; her vector addition, by itself, was faultless. It is more that her naive physics and classroom physics stood side by side but unrelated, and in this instance she exercised her naive physics. (ibid. p.59)

A further misconception held by the girls was that when an object is kicked into motion it will stop after a while. We live in a world of friction and although Newton's First Law states that when a kick sets an object in motion it will keep moving forever at a constant velocity as long as no other force acts on it, we apparently see the contrary. We know that, as there is friction, this object will always slow down until it stops. Amrit thought the machine was cheating when she first kicked and the turtle did not stop after it had got to a certain point. When the comment was made to Angela that she always brought friction into her model, even when it was stated that there was no friction, she replied: 'In my head, yes, I think it is because of what actually happens. It is difficult to think of anything that does not involve friction'. Because of these unseen frictional forces, it is assumed that where there is a constant velocity, there must be a constantly acting force producing it. It was observed that students erroneously extended the belief that constant motion implies constant force into the frictionless situation (Clement, 1977). Similarly, most of our students when considering the forces acting on a projectile talked about the force of gravity acting down and another force, variously described as: 'the force that's keeping it up', 'the force it was given originally' and 'the force acting along the curve'.

Experimenting with the Newton Microworld gave the girls the opportunity both to investigate what really happens in a frictionless world and to discover Newton's Laws for themselves. This produced a lot of surprises, disbelief, and at times confusion. Initially Nazneen talked about 'feeling muddled'; 'I thought I got everything wrong', she said. As we shall see below, although Angela could see that the motion of a particle in a circle depended on a force acting towards the centre of the circle, intuitively it did not feel correct. 'Yes, one [force] towards the centre of the circle. Whereas my reasoning would say that there was another one pushing it to keep it from going towards the centre. That is what you would think anyway. Most people would think that there was one keeping it in the circle.'

Some illustrative descriptions of their work is given in the following case studies of two of the girls.

Angela

Angela is studying A-level mathematics as well as music and history. She is in a school where the computer is incorporated into the mathematics learning and the students have a weekly Logo session. Angela is a lively student who enjoys her mathematics and feels fairly confident in her ability. She is the only girl in her mathematics class and so normally works on the computer together with a boy. When she talked about why she came to the extra sessions and what she enjoyed about them, her comment was: 'It is really good being able to work on the computer all by myself for a bit. I never thought I could have worked that [orbital motion] one out'.

Angela had already used the Newton Microworld but she still retained the naive physics view. She tried to apply her classroom physics when asked to describe the mechanics of a ball but failed to remember the rules. 'The force down there and then, I can't remember. Then there will be the force going there, due to its motion, going that way, or is it that way? I can't remember.' When asked to think about it rather than try to remember, her naive, intuitive view came forward. 'There is gravity pulling down, that is one I am sure of. Then there is a horizontal one, because the drift of the motion is horizontal, and there is a tangential one, because that is the general direction of motion.' Finally 'there is a one [force] going up, because it is moving up. Otherwise it would just move straight across.' Here she is trying to end up with a resulting force in the direction of motion. When she used the projectile program she found, 'when I asked it to print forces it just printed the gravitational force and not the force I kicked it with. I was expecting both of them to show.' She knew that at the top of a projectile's flight the vertical but not the horizontal velocity was zero. When asked about the forces at this point, Angela said 'One there [gravity], one there [horizontal] and there is one there [vertically up] which sort of cancels gravity out and so it doesn't move up or down, just across.' When she started to think about what was the force acting vertically up, she knew there was no string or wire there to hold it up. Her naive and classroom physics came into conflict and she started to panic. This is where the microworld work was so important. She could face her confusions and spend time studying the projectile motion with the turtle. This helped her to understand the conflict and rediscover Newton's Laws in practice.

Angela spent a lot of her time on the computer modelling the motion of a particle moving in an orbit under the action of a central force. While talking about forces on a body moving in a circular motion she was convinced: 'There is a force going towards the centre of the circle, that is what keeps it in the circle. But there is another force that is related to that force, there is one that is pulling out. It is not straight out like that, then it would just stay stationary. It must be either that way or that, I think it more likely that way, or very slightly off.' When she modelled this on the computer 'the turtle went crazy'. She worked for several hours modelling the motion using the Newton Microworld. It was some time before she would consider only a force towards the centre of the circle. She was really excited and pleased when she succeeded in writing a procedure to move the particle in an orbit. With this procedure she was then able to test out what would happen when the central force was removed. Even when Angela saw the particle moving round in a circular orbit she was still expecting there to be two forces acting on it. Time spent printing out force diagrams and then testing out her circle program helped her clarify the ideas. When talking about circular motion after her sessions on the computer Angela explained what she had learnt from this time:

> *Angela:* Now I know there is a force acting, there towards the centre and the particle is travelling around it. Acceleration is towards the centre of the circle whereas actual motion is at right angles to the acceleration. And there is a force attracting the ball to the centre, which is why the ball is accelerating. If that is taken away the ball carries on travelling at the tangent at which it was travelling before. Which I didn't think before. I thought maybe it would curve out. The computer has helped me think that through.
>
> *Interviewer:* How did you feel when you saw the ball travelling out at a straight line like that?
>
> *Angela:* I don't know, I think I half expected it. It triggered off some memory about doing mathematics at school when we did circular motion, but I hadn't done it for months.

Angela then extended her work to look at orbital motion. She used this to investigate the relationship between the force of attraction on a body in orbit and the distance from the attracting body. She left the session feeling confident that she had worked things out for herself. She wanted to continue the work. 'There is no reason why I can't do this every Friday. It will help relieve my memory. The whole point of doing Logo is to get this clear.'

Sadia

Sadia is studying chemistry, biology and mathematics at A-level. As we have said, she finds the mathematics harder than the other two subjects and uses the word 'frightening' to describe her feelings about it. She had never touched a computer before and when she came to the first session was nervous, not knowing what to expect or what was expected of her. She worked on her own machine, but alongside a friend, and there were often discussion, comments and ideas running between them. Sadia was happy to tell us how she was feeling about the work and got into the habit of typing comments about her progress into the computer. These could be her predictions about how the turtle would move in response to her commands or exclamations at unexpected behaviour: 'why is the turtle moving so fast?' and 'Hello, hello, hello. Why doesn't it exist??? Help!'

Sadia had already studied the motion of a projectile in her applied mathematics classes. She knew and felt she understood its properties. When she discussed it in more detail, her naive physics took over. At any point, she felt there must be a force, other than gravity, acting on the body: 'there must be something keeping it up'. At the top, the two forces 'must be equal, because it is stationary. Therefore the two forces must cancel each other out.' She did not distinguish between velocity and speed: 'both mean the same'.

Sadia started her work on the computer by investigating the effect of a kick on a turtle moving in a particular direction. She looked at the effect when the kick was in the same, a different or the opposite direction to that of the motion. The results were surprising: 'I don't know what I was expecting'. When the kick was in the opposite direction, then she predicted that the turtle would move off in that direction and 'felt very confused' when it continued to move in its original direction. This led her to carry on with her exploration of the effect of a kick and to rediscover Newton's Laws for herself. Sadia then worked on a projectile procedure, editing it where necessary, to explore the effect of gravity on the x and y components of the speed. Here she discovered that the x-speed was unaffected and the y-speed was changed by steady amounts. Having adjusted to this discovery, she set herself a new task, to work on what she called a 'sideways projectile problem'. Here 'gravity' acted parallel to the x-axis instead of parallel to the y-axis. This was not a straightforward task: 'why is the turtle moving so fast?' and 'how do I make the path shorter?' However after some time she said: 'I've got it moving in the sideways position' and became very excited with this achievement. She started to explain the procedure to Amrit although full of self doubt: 'Help, I can't remember now, I bet it doesn't work.' In the process of explaining, the principle became clear enough for her to say to Amrit: 'Oh I see now, it's easy.' When Sadia reflected on what she had learnt by using the computer and in particular in writing her own 'sideways' procedure it was clear she had picked up a great deal about the motion of a body under gravity. Now she recognized that only the y-speed changes and this is by a constant amount. She had also investigated the relationship between the speed of projection of the particle and its range: 'If I give it a different speed to begin with, that will determine how far it will go. Before I didn't think it was important what distance it went.' She now accepted that, for a projectile, only the force of gravity was acting.

However, intuitive ideas of physics still remained and statements such as, 'the other force, the one that is actually making it move in a curve, 'cos if there was only this force [gravity] it would fall down', did venture back into her conversation. This led to a contradiction for she was confident that the horizontal speed stayed constant and she was also by now sure that applying a force leads to a change in speed in the direction of the force; hence there could not be any force acting 'along the curve'. The recognition of this contradiction was her parting thought. More time was needed for her to check out and reinforce all her ideas.

For someone who had no experience of computers, working on the machine caused Sadia very little anxiety. When the turtle behaved in an unexpected way, the frustration was with her mathematical model of the situation rather than with the machine. Her excitement at turning the projectile motion on its side was partly her pleasure in forming a good working mathematical model of the process but also at writing a procedure herself 'from scratch'. She left feeling she could continue to work on the computer on her own. In her final interview Sadia spoke with much more confidence and conviction about the mechanics of a moving body. This contrasted with her uncertainty at the beginning.

Conclusion

In the light of this work we would like to put forward the following general comments about the girls we worked with.

Prior to this experience most of the girls had tended to think of computers as in the boys' domain, or not available to them. When they were able to work in a supportive environment, however, the presence of hardware was not inhibiting, rather the computers were seen as a tool to help sort out ideas. They were keen to take advantage of an opportunity to use computers in ways that would help their mathematics. In particular the opportunity to use a dynamic modelling tool was useful to those who had not studied physics. It gave them the opportunity to move away from rote and build up an understanding based on simulated experience.

The girls usually worked co-operatively, even if on individual machines, and the process of discussing or explaining led to better understanding. However, they sometimes needed the chance to work on their own, rather than in pairs, on the computer. They were emotionally as well as intellectually involved and frequently talked about their feelings of excitement, frustration and fun. They did not mind expressing their misconceptions, since there were ways of coping with any contradictions that arose. They went away with a sense of achievement and new confidence, and wanting to do more work on the computer.

We want to thank all the students who took part in the study, and we hope they will continue working and experimenting with computers.

References

Burns, S. (1988), 'Learning mechanics with Logo: a study of sixth-form students experimenting in a Newtonian Microworld', unpublished MSc dissertation. Institute of Education, University of London.

Clement, J. (1977), 'Intuitive, non-Newtonian views of motion', unpublished mimeo.

di Sessa, A. (1982), 'Unlearning Aristotelian physics: a study of knowledge-based learning', *Cognitive Science,* Vol.6, pp.37-75.

Gem Software, *The Newton Microworld for Nimbus Computers.* Gem Software, 58 Parklands Avenue, Lillington, Leamington Spa, Warwickshire.

Mathematics Association (1985), Proceedings of the A-level Conference held at Ware, *Draft Interim Report No.2,* April.

Turkle, S. (1984), *The Second Self: computers and the human spirit.* London: Grenada.

Geometrical Thinking and Logo: do girls have more to gain?

Richard Noss

Sutherland and Hoyles in their chapter above have provided convincing evidence that the programming styles of boys and girls differ significantly. One way of viewing this difference is to consider the ways in which boys and girls interact with the culture generated by the Logo environment, that is, to think of the computer as influencing not only the setting within the classroom but also the culture within which children do and learn. Of course, the development of a genuinely pervasive computer culture of the kind suggested by Papert is, for the most part, some way in the future. Nevertheless, there are enough pointers — especially within restricted subcultures (see, for example, Turkle, 1985) or for the learning disabled (Weir, 1987) — to suggest that the phenomenon might be interpreted culturally, i.e. as an interaction between existing cognitive and affective styles and preferences, and new possibilities opened by the computer.

If such an interpretation makes sense at all, then it might be expected that — at least in the long run — the computer's presence might bring about changes, not only in the *way* mathematics is learned, but in the content of *what* is learned. Further, if boys and girls' programming styles consistently show differences, then we might possibly expect differences to show up in what they take away with them from their Logo work in terms of understanding. Unfortunately, investigating such a hypothesis is problematic. As Sutherland and Hoyles point out in their criticism of Carmichael et al.'s study, it is not at all obvious what to measure; worse still, it is all too easy to collect data which prejudges the very issues under investigation. The difficulty is compounded since experience of the computer within the classroom is limited, and research so far has been sparse. So the question is, what kind of learning might it be reasonable to look for — and how?

There have been a number of studies which have looked for 'effects' from the Logo environment (see, for example, Pea and Kurland, 1984). In the main, these have hypothesized that experience with Logo will lead to an enhanced understanding of some concept or heuristic. Perhaps unsurprisingly, these have often yielded at best ambiguous results. If it is possible to generalize at all from the results of research into this question, it is that transfer is — as in other domains — extremely difficult to achieve; and when it does appear, it is as a result of a conscious and concerted pedagogical strategy, not by accident.

It seems to me that looking for transfer effect *per se* is not helpful. Rather than look for the effect of learning Logo on some mathematical concept which is subsequently (or concurrently) taught, it makes more sense to ask whether the experience of learning Logo actually influences underlying mathematical conceptions which are not taught at all. If Papert is right in considering Logo programming as providing external representations (like mental gears) on which to base learning, then it might be the case that precisely those facilities which are not taught, but which are 'merely' acquired through participation within the general learning culture inside and outside school, are those that are influenced.

Papert (1980) discusses the relative difficulties experienced by children in discovering various Piagetian concepts which are, in the main, 'acquired' without teaching. He argues that the effect of the spread of an environment in which the computer is commonplace, will ultimately influence and even reverse the developmental order in which children acquire or discover fundamental concepts:

> If children grow up surrounded by computers and a computational culture, it seems quite plausible to me that they will find such problems as forming families of beads perfectly concrete and be able to carry them out as early as they discover the conservation of number. And if computers become really important in their lives, they may develop the computational concepts even earlier than the numerical, thereby reversing what has appeared to be a universal of cognitive development. (Papert, 1980, p.995)

It is difficult to clarify just what might be the specific action of the computer experience on children's underlying conceptions. One way of thinking about such an action is to take up Papert's interesting suggestion that underlying cognitive structures — Piaget's *groupements* — are best understood as 'microworlds' (schemas, frames) created by the child to enable her to make sense of concepts which are too deep to be acquired in one unit. Some general evidence on this question has been supplied by Lawler (1985) who offers evidence that a young child remote from the formal stage 'developed through programming experience characteristics of thought related to those typical of formal thinkers' (Lawler, 1985, p.74).

Viewed from this essentially Piagetian perspective, other fundamental misconceptions become susceptible to explanation. Consider, for example, the finding of Hart (1980) that some 58 per cent of 11-year old pupils could not conserve length; explicitly, they were unable to judge which of two lines (one oblique and one horizontal with their end points aligned) was the longer (see Figure 1). This finding is robust, and has been confirmed in a number of studies since then, as well as in a number of different cultural settings. How should we interpret this result? In the first place it implies that the ways in which adults' understand the concept of length is not shared by all children. At first sight this is remarkable, given that 'length' is a common idea in everyday use. In fact, this very observation provides at least one possible explanation. For, as Minsky (1987) argues convincingly, some of the most important developments in individual cognition occur not by learning new facts, but by developing ways of managing existing ones[1]. Length is a concept with a wide range of meanings within the culture, some of which actually *do* ignore or at least downplay the orientation. The problem for the learner is to know when it is appropriate to pay attention to what.

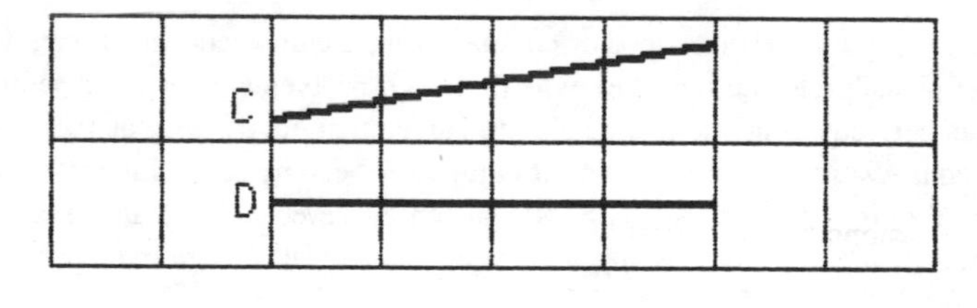

Line C is longer ____________
Line D is longer ____________
C and D are the same length ____________
You cannot tell ____________

Figure 1: Length conservation problem (from Hart, 1980)

1 Minsky attributes this insight, which he calls 'Papert's Principle', to Seymour Papert.

Of course, if the problem was phrased in context (e.g. is it shorter to run across the field diagonally, or to run round the touch-line?) we may safely assume that 100 per cent of children much younger than those in Hart's study would answer correctly (see, for example, the study by Jean Lave and her colleagues, 1985). In other words, the context in which the problem takes place is crucial in handing over control to the appropriate 'agents', in Minsky's terms, for the problem at hand.

It is a facet of the Logo environment that the learner's attention is inevitably concentrated on certain key components of the environment, and it is in this sense that it makes sense to talk of a microworld based on a specific mathematical conceptual field (for a discussion of this issue see Hoyles and Noss, 1987). This focusing aspect of the computational environment should not be underestimated; in a 'normal' classroom, it forms a crucial — perhaps *the* crucial — part of a teacher's rôle. How often does it seem as if 'all I had to do was to get her to read the question'? In such cases, the student's problem is to have failed to locate what is important, often a failure to engage in the kind of decontextualization that is so often a part of the school mathematics curriculum.

In a Logo environment, the Turtle-geometric microworld is one in which the idea of length (in the form of 'going forward') and of angle (in the form of turning right and left) are 'the things you have commands to change' (Weir, 1987). So it is exactly the ideas of length as a distance, and angle as a dynamic turn, which we might expect to develop as a result of using Logo.

★ ★ ★

At this point, I would like to draw on some empirical results from a study which has been reported in detail elsewhere (see Noss, 1987). This study was not ostensibly to investigate gender differences; it aimed to compare the responses on a set of geometrical problems of children who had studied Logo for one year, in comparison to a group who had no computing experience. Given my contention that we are dealing here with an essentially cultural phenomenon, it is perhaps useful to outline in a little detail, what the 'Logo' children's activities involved.

Eighty-four pupils in four classes, aged between 8 and 11 years (in four different primary schools), programmed in pairs for one year. Their programming work was largely unstructured, in that they were mainly free to pursue their own goals with the help and intervention of their class teacher when required. No explicit attempt was made to help the pupils to synthesize their Logo work with more formal geometrical knowledge. By the time of the experiment reported here, each child had undertaken an average of some 30 hours programming, consisting of approximately 75 minutes per week in pairs. (Full details of the children's Logo activities can be found in Noss, 1985.)

The experiment was designed to probe the concepts underlying the children's notion of length and angle, rather than to test for 'knowledge' of mathematical understanding *about* those concepts. Accordingly, there were three categories of problems for the concept of length, and three for angle. For example, the category of length 'conservation' consisted, among others, of the problem used in Hart's (1980) study discussed above. Other length categories were 'combination' (seeing that a given length is composed of two sub-lengths), and 'measurement' (concentrating on the unit of measure as well as the value of the length). For angle, the categories were: 'conservation' (seeing that the size of an angle — analogously to the length conservation category — is invariant under rotation and length of rays); 'right-angle conservation' (a specialization of the previous category); and 'measurement' (ordering of angles by size). The test was piloted and given to the children at the end of the academic year.

The essentially illuminative nature of the study as a whole precluded a strict experimental design. This should be borne in mind when interpreting the results. Nevertheless, there do exist statistical techniques which are sensitive to the kind of interactions which might exist in data of this kind: the reader is referred to Noss (1987) for details of the technique employed. Since the main interest here is on using the findings of the study to focus thinking about the issues rather than to re-present results reported elsewhere, I shall concentrate on just two illustrative findings, both concerned with the idea of conservation in a sense analogous to, but not identical with, Piaget's use of the term.

I begin with Hart's (1980) familiar conservation of length item already referred to above. As I pointed out earlier, the facility level of this item in Hart's study was extremely low (42 per cent for the first-year secondary students). Not surprisingly, especially when the age-difference of between one and three years is taken into account, a similarly low facility level was found in the study I am reporting here. However, there was a trend in favour of the Logo group who scored around 39 per cent, while their non-Logo conterparts scored some 10 per cent lower. When the results were broken down by sex, much more interesting findings emerged. In both the Logo and the non-Logo groups, the boys outperformed the girls by a significant margin. There was, however, an indication that this gap had been narrowed in the case of the Logo groups: for the non-computer groups, the girls' score was just under half that of the boys, whereas for the Logo groups the gap had narrowed to two-thirds.

The interpretation of these results is somewhat hampered by the fact that Hart's results are not differentiated by sex, so it is impossible to gauge the full extent of this trend, which in any case is rather weak. However, before commenting further, I should like to refer to a second aspect of conservation which was investigated, that of conservation of angle. The item I shall concentrate on is shown in Figure 2.

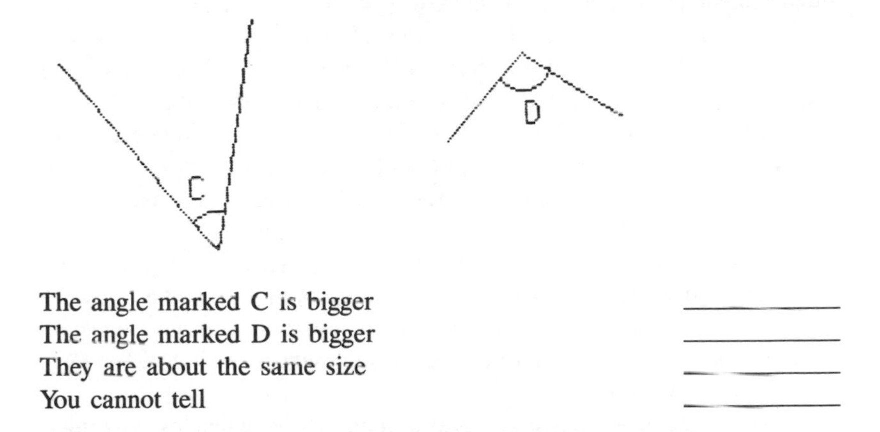

C
D
The angle marked C is bigger
The angle marked D is bigger
They are about the same size
You cannot tell

Figure 2: Angle conservation problem

Here there were two aspects of the data which differ from those for length. In the first place, the difference in performance between the Logo and non-Logo groups was statistically significant in favour of the Logo groups. Secondly, when the results were analyzed by sex, a pattern broadly convergent with those for length emerged — but somewhat more strongly. In the non-Logo groups, the boys scored consistently higher than girls; in the Logo groups, this gap had effectively disappeared.

There are a number of points to be made in connection with these results. In the first place, similar findings were recorded in five of the six categories outlined earlier (the exception being that of length 'combination'). So the findings seem to be indications of a trend — no more than a trend — which might need some explanation. Secondly, it is clear that the study contains within it limitations which restrict the scope for generalizability: the lack of data concerning the children's initial conceptions at the outset of the study, and the impossibility of a more rigorous research design are just two. Finally, it is clear that success or failure on written tasks is only one measure — and not a particularly sensitive one — of performance; while subsequent research has adopted more sensitive methodologies, these did not form part of the study reported here.

Let us first consider the result that the findings for the conception of length conservation were less marked than those for angle. It seems that this difference might throw some light on the mechanism at work in general. Papert et al. (1979) found something similar (again giving children a paper-and-pencil test); they conjectured that new knowledge acquired in the Logo environment has to 'compete' with existing knowledge, and that the amount of time required to displace it would depend on how firmly rooted it was. It is reasonably easy to agree with Papert's conjecture that knowledge about angles would be more easy to displace than knowledge about length. After all, the concept of length and its associated invariants (longer, shorter, etc.) are very much a part of the youngest child's experience; her environment is rich in the kind of 'intellectual material' (Mellin-Olsen, 1987) that underlies the notion of length. The same cannot be said of angle, at least not in the static form of the usual mathematical conception. That the situation might be different in a turtle-geometric context, where the movement of the turtle corresponds to the child's intuitive notions derived from her own body movements, is of course part of Papert's argument in favour of the turtle approach to geometry.

If we accept the explanation based on competitive agencies, this might go some way towards explaining the differential effect in favour of the girls. There is, of course, plenty of evidence of some (perhaps small) differences between girls and boys on spatial and geometric tasks, at least on some tasks with some methodologies. Less contentiously, there are the evident socio-cultural tendencies which militate in favour of boys playing with constructive toys and participating in activities which encourage spatial awareness, and a corresponding tendency against girls' participation in similar activities. Moreover, there is evidence that such 'deficits' as may exist are remediable through appropriate activities (Sherman, 1980).

If this is the case, then it might follow that Logo could act as some kind of remediating experience. What is interesting here is that such learning takes place without explicit teaching; that is — at least this is the conjecture — that by enlarging the culture within which ideas of movement and turn take place, it might be that the varieties of meanings of concepts such as 'longer' and 'bigger' are sufficiently enriched to bring about some changes in children's underlying conceptions. Much in this argument hinges on the assumption that Logo provides (or can provide, given the appropriate setting and pedagogical input) just such a culture. There is some evidence on this; children working in pairs on Logo *do* tend to focus their discussion on the task in hand, and there is plenty of corroborative data that their discourse is rich in the kind of content we are considering here (see Hoyles and Sutherland, 1986).

Of course, it is somewhat presumptuous to talk of changing children's culture by introducing a couple of computers into a classroom. Bringing about a computer-rich culture entails, of course, much more than this: not simply the provision of sufficient hardware (say, two computers per child), but the construction and development of associated language, expression and media (not to mention an appropriately revolutionized curriculum in school). What little information exists about this kind of situation suggests that in a computer-rich school environment, at least, genuine changes in the social construction of knowledge actually do take place[2].

The implication is that much more needs to be known about the kinds of activities which characterize such an environment. It is clear that Logo is not alone in presenting possibilities for informal learning to take place. Indeed, one of the most promising developments currently taking place is the construction of a new generation of computer software which offers the learner the possibility of operating at an informal intuitive level, while the machine constructs formal mathematically precise models behind the scenes. (Contrast this with Logo: here the learner writes programs, i.e. constructs formal symbolic code, and receives visual, intuitive feedback.) It is much too soon to guess what the implications of such software might be; nevertheless, the traditional boundaries between programming and other computer applications are breaking down. Already, the new generations of Logo are offering integrated programmable environments (such as word-processors and control technology software), which mean that the 'user' is programming for a purpose rather than simply for programming's sake. Such developments might prove particularly rich for girls[3].

There is emerging evidence that the computer is potentially capable of

2 Seymour Papert and his colleagues at the Massachusetts Institute of Technology have recently set up a computer-rich environment in a Boston elementary school.

3 Taylor (1986), for example, suggests that girls are more inclined to become involved in Lego construction if the activities are incorporated into a story format.

interacting with pupils' existing and developing mathematical knowledge in ways which expand their understandings. Hoyles and Noss (1989) have illustrated how the computer offers children the opportunity to succeed at mathematical tasks which they would be unable to do with pencil and paper. Similarly, Noss (1986), and Sutherland (1988) have illustrated how children's experience of programming seems to be effective in generating underlying algebraic conceptual frameworks which can act as the foundation for more formal algebraic learning. As yet, there is little information on the influence, if any, of gender on this and similar processes; research in this area could prove a fruitful source of insight, not only into the specific ways in which girls and boys may differ in what they take away with them from their computer-experiences, but into the interaction between the computer, learning styles, and the development of mathematical conceptions in general.

References

Hart, K. (1980), *Secondary School Children's Understanding of Mathematics. A report of the mathematics component of the CSMS programme.* Chelsea College, University of London.

Hoyles, C. and Sutherland, R. (1986), 'Peer interaction in a programming environment', *Proceedings of the Tenth International Conference for the Psychology of Mathematics Education*, pp.354-359. London.

Hoyles, C and Noss, R. (1987), 'Synthesizing mathematical conceptions and their formalization through the construction of a Logo-based school mathematics curriculum', *International Journal of Mathematics Education in Science and Technology.* Vol.18., No.4, pp.581-595.

________ (1989), 'The computer as a catalyst in children's proportion strategies', *Journal of Mathematical Behaviour,* Vol.5 (in press).

Lave, J., Murtaugh, M. and de la Rocha, O. (1985), 'The dialectic of arithmetic in grocery shopping' in B. Rogoff, and J. Lave (eds), *Everyday Cognition: its development in social context.* Cambridge, MA: Harvard University Press, pp.67-94.

Lawler, R. (1985), *Computer Experience and Cognitive Development.* Chichester UK: Ellis Horwood.

Mellin-Olsen, S. (1987), *The Politics of Mathematics Education.* Holland: Reidel.

Minsky, M. (1986), *The Society of Mind.* Cambridge, MA: MIT Press.

Noss, R. (1985), *Creating a Mathematical Environment through Programming: a study of young children learning Logo.* London: Institute of Education, University of London.

________ (1986), 'Constructing a conceptual framework for elementary algebra through Logo programming', *Educational Studies in Mathematics,* Vol.17, No.4, pp.335-357.

________ (1987), 'Children's learning of geometrical concepts through Logo', *Journal for Research in Mathematics Education,* Vol.18, No.5, pp.343-362.

Papert, S., Watt, D., diSessa, A. and Weir S. (1979), *Final Report of the Brookline Logo Project, Part 2.* AI Memo No.545, MIT, Cambridge, MA.

Papert, S. (1980), 'Redefining childhood: the computer presence as an experiment in developmental psychology', *Proceedings of IFIP Conference.* North-Holland.

Pea, R. D. and Kurland, D. M. (1984), 'On the cognitive effects of learning computer programming', *New Ideas in Psychology,* Vol.2, No.2.

Sherman, J. (1980), 'Mathematics, spatial visualization and related factors: changes in girls and boys, grades 8-11', *Journal of Education Psychology,* Vol.72, pp.476-482.

Sutherland, R. (1988), 'A longitudinal study of the development of pupils' algebraic thinking in a Logo environment'. Unpublished doctoral thesis, Institute of Education,
University of London.

Taylor, H. (1986), 'Experience with a primary school implementing an equal opportunity enquiry', in L. Burton (ed.) *Girls into Maths Can Go,* London: Holt Rheinhard and Winston, pp.156-62.

Turkle, S. (1985), *The Second Self: computers and the human spirit.* New York: Simon and Schuster.

Weir, S. (1987), *Cultivating Minds: a logo casebook.* New York: Harper and Row.

FORUM

for the discussion of new trends in education

FORUM, founded in 1958, is an independent journal addressed to progressive classroom teachers, heads and administrators, as well as to parents interested in understanding more about new developments and trends in education.

FORUM is run by teachers: the editorial board is drawn from infant, junior and comprehensive schools, adult and community education, administration and teacher education. It is an entirely independent journal, having no connection with any established organization or institution.

FORUM keeps close to the classroom, but it is also alive to the issues behind the news and behind new developments in education. It has been in the forefront of the move towards comprehensive education and towards mixed ability grouping in primary and secondary schools — trends which FORUM pioneered.

Articles regularly discuss the content and methods of education: new teaching methods, classroom organization, curriculum, multicultural education, assessment, profiles, classroom and school management. Government policy is regularly analysed and assessed.

Publication dates: September, January, May.

BEDFORD WAY PAPERS

educational studies and related areas

THE NATIONAL CURRICULUM (BWP 33)

Denis Lawton and Clyde Chitty (eds.)

A proposal for a national curriculum for all maintained and voluntary schools in England and Wales has been incorporated into the Education Reform Bill currently before Parliament. By the 1990s (if the Bill receives the Royal Assent) the new curriculum and assessment structure can be expected to be fully in place. It will represent a revolution in national education policy, the reassumption by central government of the direction of the school curriculum it had ceded earlier in the present century and a remarkable throwing off of an ingrained suspicion of state power in education. A giant step is being taken: how successful will it be in achieving the Government's declared purpose of 'A better education — relevant to the late twentieth century and beyond — for all our children, whatever their ability . . .'?

Nine contributors to this Bedford Way Paper — all members of the teaching staff of the Institute of Education — subject the national curriculum policy to urgent criticial examination. Government ministers have presented the policy as a natural next step in an evolution that has been gathering pace over ten years or more. Some contributors here see it in quite a different light, arguing that a growing professional consensus in favour of a common curriculum based on 'areas of experience' or the 'common culture' has been rejected in favour of a subject-based curriculum which looks back to a much earlier period. In their view a 'bureaucratic' curriculum is being adopted, in essence concerned with efficiency and the need to obtain precise information through testing to demonstrate it. The contributors look in vain to the proposal for an intellectually satisfying account of the general aims of education and with concern at the changes to the character of education given in our schools which might ensue. This Bedford Way Paper will be essential reading for anyone concerned about national curriculum policy.